FREE BOOKS
www.*forgottenbooks*.org

You can read literally thousands of books for free at www.forgottenbooks.org

(please support us by visiting our web site)

Forgotten Books takes the uppermost care to preserve the entire content of the original book. However, this book has been generated from a scan of the original, and as such we cannot guarantee that it is free from errors or contains the full content of the original. But we try our best!

Truth may seem, but cannot be:
Beauty brag, but 'tis not she;
Truth and beauty buried be.

To this urn let those repair
That are either true or fair;
For these dead birds sigh a prayer.

Bacon

MODEL ENGINE-MAKING

In Theory and Practice.

BY

J. POCOCK.

*WITH OVER ONE HUNDRED ILLUSTRATIONS,
DRAWN BY THE AUTHOR.*

LONDON:
SWAN SONNENSCHEIN & CO.,
PATERNOSTER SQUARE.
1888.

MODEL STEAM ENGINES.

Every description, from the smallest to powerful Engines, of the very best workmanship, at the lowest possible prices. LEE'S NEW ENLARGED and REVISED ILLUSTRATED CATALOGUE contains Illustrations and Descriptions of Engines from the smallest to 15 H.-P. Paddle, and Screw Steamboats from 1s. to 37 guineas; Sailing Yachts, Castings of Engines, Tools, Lathes, Shaping Machines, Foot Motors, Chucks, Parts of Engines, Anchors, Guns, Boiler Fittings, &c., Electric Machines, &c. To Amateurs the most useful book extant, and largest and best Catalogue in the Trade, 64 pages, 110 Illustrations, price 6d. post free.

ENGINE CASTINGS.

LEE'S WORLD-FAMED BRASS CASTINGS, with his improved "Tenon" or Chuck Pieces, on all parts requiring to be turned, are supplying a want long felt, being easily turned even by the most unskilled amateur. Complete sets for Horizontal Slide-valve Engine, $\frac{1}{2}$ in. bore, 1 in. stroke, 2s., post free, 2s. 3d. ; $\frac{3}{4}$ in. bore, 1$\frac{1}{2}$ in. stroke, 3s., post free, 3s. 6d. ; 1 in. bore, 2 in. stroke, 4s. 6d., post free, 5s. ; 1$\frac{1}{4}$ in. bore, 2$\frac{1}{2}$ in. stroke, 11s. 6d., post free, 12s. 6d. ; by Parcels Post. All sizes up to 6 in. stroke at equally low prices. Each Casting has Lee's improved "Tenon," without which model Castings are practically useless.

TO AMATEURS.

R. A. LEE is now supplying all his best HORIZONTAL ENGINE CASTINGS, from 1 in. bore, with the Steam and Exhaust Ports accurately cast in the Cylinders, free of charge, thus saving an immense amount of work in fitting up and ensuring accuracy in one of the most important parts of the Engine.

PHOTOGRAPHIC LENSES *at Marvellously Low Prices.*

THE EXCELLENS.

Single Achromatic, to cover 5 by 4, 7s. 6d. ; ditto, half-plate, 10s. 6d. Rapid Rectilinear Lens, with set of Waterhouse Diaphragms for Instantaneous work, to cover 5 by 4, 21s. ; half-plate, ditto, 30s. Each of these rapid Rectilinear Lenses will cover the size larger than it is stated to be—*i.e.*, $\frac{1}{2}$-plate will cover 1-1 size. All post-free.

These Lenses are guaranteed of the highest quality, and cash will be returned if not satisfactory. Illustrated Catalogue gratis on receipt of addressed postal wrapper.

To be obtained only of **ROBERT A. LEE,** Manufacturing Optician. Works :—76, 77 and 78a High Holborn, London, W.C. (directly opposite Inns of Court Hotel).
Please state where you saw this advertisement.

AWARDED PRIZE MEDAL INTERNATIONAL INVENTIONS EXHIBITION.
Lee's Patent Electro-Motor and Battery for Driving Lathes, &c.
Half H.P. for one-third the price of a Gas Engine. Batteries for Electric Lighting.
Send for Prospectus gratis on receipt of addressed postal Wrapper.

ROBERT A. LEE, ENGINEER,
76, 77 & 78a, High Holborn, London, W.C.

WORKING MODEL STEAM ENGINES
Largest Stock in the World.

JOHN BATEMAN & COMPY.,
Engineering Modellers, &c.

Originators of Castings for the Construction of Model Steam Engines, Lathes, &c., &c., by Amateurs.

BATEMAN & COMPANY'S "MUSEUM OF MODELS" pronounced by the leading press of the day to be one of the *most interesting* sights in London, comprises five houses, and has on view the Largest Stock of Working Models in the World, from 2/- up to £500 each.

VERTICAL STEAM ENGINE MODEL.
(Works beautifully.)

Complete with Boiler, Furnace, Brass Cylinder, &c., &c., 2/-; Post free, 2/3.

LOCOMOTIVE RAILWAY ENGINE.
(Works perfectly.)

7¼ in. long, 2 Brass Cylinders, &c., &c., 5/6 complete. Postage 5d. extra.

CASTINGS

Of every conceivable type of HORIZONTAL, BEAM and OTHER MODELS in gradual running sizes from ¼ in. to 4 in. bore, from 2/6 per complete set.

MARINE and LAUNCH ENGINES (single and double), ¼ to 4 in. bore.

CASTINGS FOR LOCOMOTIVE ENGINES,
From the simple Oscillating Cylinders to the most perfect Scale Models.

CASTINGS OF LATHES,
From 2¼ Centre Back and Single Geared up to 6 in.

❋ Every Requisite for Amateur Engineers. ❋

See J. BATEMAN & COMPANY'S interesting MECHANICAL PAMPHLET and CATALOGUE, 100 Illustrations, giving fullest details of the above, 6d., post free. Address:—

JOHN BATEMAN & COMPY.,
"Museum of Models,"

HIGH HOLBORN, LONDON;
Also at the
"ORIGINAL" MODEL DOCKYARD,
53, FLEET STREET, & 104, STRAND.

Model Steam and Sailing Ships, Yachts, Blocks and Fittings.—New Pamphlet, fully illustrated, 4d., as above.

PREFACE.

THE Steam Engine in one or other of its many forms is a sufficiently familiar object; not equally familiar, I venture to think, the knowledge of its manner of working. Yet surely in these days of scientific and technical education it would be well that at least the younger members of the community should have some clear perception of the method in which steam exerts the force which has become so necessary to the exigencies of daily modern life.

The want of some practical book on Model Engine-Making has during the last few years been made evident by the number of queries upon this subject which have appeared in various journals more or less devoted to the interests of mechanical science.

Much of the desired information might no doubt be gathered from the columns of these papers, notably, for instance, from the back numbers of *The English Mechanic;* but this involves a troublesome search and a comparison of sometimes diverse directions.

These and kindred considerations have induced me to revise and enlarge a series of papers on Model Engine-Making, which originally appeared in a magazine chiefly addressing itself to the interests of amateur workers.

Wishing these papers to be useful to as large a body of amateurs as possible, I have thought it best to give many details for the sake of beginners, which may not be needed by those who have already become practised workmen.

The patterns in which Model Engines are made are very numerous, but they will usually be found to be only varieties of one or other of the types mentioned in this book; and any one who has constructed the models here described will, I believe, find little difficulty with any other pattern, whether it be that of a model or a small power engine.

J. POCOCK.

HARLESDEN, N.W.

CONTENTS.

CHAP.		PAGE
I.	Tools and Castings	1
II.	Oscillating Cylinder Engine	12
III.	Oscillating Cylinder Locomotive	30
IV.	Horizontal Slide-Valve Engine	42
V.	Horizontal Slide-Valve Engine (*continued*)	70
VI.	Vertical Slide-Valve Engine	91
VII.	Launch Engine	99
VIII.	Marine Engine	109
IX.	Locomotive Engine	122
X.	Model Boilers	141
XI.	Boiler Fittings, etc.	151
XII.	Notes	168
	Index	177

MODEL ENGINE-MAKING.

CHAPTER I.

INTRODUCTION.

TOOLS AND CASTINGS.

THERE may be those who, never having tried their hands at model engine-making, are ready to maintain that such work is a mere waste of time; but, as a matter of fact, there are few leisure occupations better calculated to train the hand, eye, and mind all at one and the same time, and to such good effect. Not only does it afford a most thorough training in all such descriptions of mechanical work as lie within the compass of an amateur's powers, as filing, drilling, turning, fitting, screw-making, etc., but a considerable amount of knowledge is also acquired as to the theory and practical application of steam—the most important power of the present day. Nor must it be forgotten that James Watt himself, although he had previously by means of a Papin's digester and a syringe made a few experiments, might nevertheless have failed to turn his attention seriously to the improvement of the steam-engine, had it not been for the model engine placed in his hands for repairs by the authorities of the College of Glasgow; and while we continue to obtain from the steam-engine as at present only a small proportion of the energy theoretically contained in the fuel consumed under the boiler, who can say but that the model-maker of to-day may be a second James Watt to-morrow?

Although many of the so-called model engines—as, for example, those described in Chaps. II. and III.—have no large and useful archetypes, still the construction of these will prove a valuable stepping-stone to the more advanced models described in later chapters; and in order to impart as much interest as possible to the work of *Model Engine-Making*, I have in many cases given the proportions of the more important parts as they would be in full-sized engines.

Before however we take in hand the actual construction of a model engine, it will be convenient to devote a preliminary chapter to the consideration of the tools required and the castings to be used.

First, as to the tools needed, these will be such as are to be found in the workshop of any amateur who has taken up metal-turning. It is sometimes stated that, provided the cylinder is bought ready bored, a lathe is not absolutely necessary. Perhaps not *absolutely*, but without it much additional time and patience will be demanded, and the results both in appearance and working will certainly be far less satisfactory. I would therefore by no means recommend any one to undertake model engine-making, unless they are provided with a lathe.

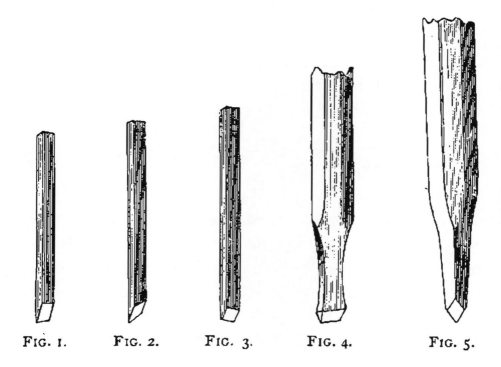

FIG. 1.   FIG. 2.   FIG. 3.   FIG. 4.   FIG. 5.

A chuck for holding small drills, and a grip-chuck holding from ¼-inch diameter upwards, will be found useful adjuncts to the lathe; and if the lathe is provided with a face-plate, so much the better.

A few small chisels of the shapes shown by Figs. 1 to 5, must be made from good square cast steel : these are for chipping out steam-ports and similar work.

A few drills from $\frac{1}{16}$-inch up to $\frac{1}{4}$-inch, of the ordinary shape, will be necessary; and D-bits of the required sizes may be made as wanted. Fig. 6 is a section of a D-bit at its cutting end, and Fig. 7 is a section taken at a little distance down the shank.

FIG. 6.   FIG. 7.

A rimer (Fig. 8) will be found useful. It is made from a piece of small steel rod, filed up to a triangular or five-sided shape, and tempered to straw colour.

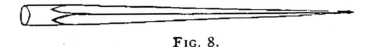

FIG. 8.

The usual gravers, round-nosed and other turning tools will be required, also one or two inside tools; and a few files, flat, half-round, and triangular, coarse, medium, and fine. A bench vice is almost as indispensable as a lathe; and an Archimedean, or some other form of hand-drill, for small drills will often be found handy in addition to the drill-chuck already named.

A small steel square, usually known as an "engineer's square," must be included in the list of tools required.

One or two scrapers will also be found useful. These may be made from broken files, the teeth being ground out for a short distance from the broken end of the file, and the end ground off straight across. These scrapers are to be used for giving the finishing touches to such work as cannot be got sufficiently level by filing.

A surface-plate must be mentioned among the tools it is very desirable to have, but it forms an expensive item, costing twelve or fifteen shillings and upwards according to size; and small work, such as slide-valves, etc., may often be got up with sufficient accuracy by using the planed lathe-bed as a testing surface

A screw-stock with dies and taps may occasionally be useful, but is not indispensable provided there is a good screw-plate among our tools.

Any other special tools required will be described as they are mentioned.

It is not, as a rule, worth while for an amateur to make his own screws, except when only one or two of some particular size are wanted; but, on the other hand, it is not always easy to procure screws suited to one's own screw-plate, unless the latter is standard thread; neither are screws of standard thread everywhere obtainable. Moreover, I must warn my readers against ordering screws, say of different lengths and the same diameter, under the belief that only one pair of taps would be required. As a matter of fact, one maker to whom I applied informed me that his screws of $\frac{3}{32}$-inch diameter and $\frac{1}{4}$-inch length were of a different thread, and consequently required a different set of taps to those screws which were $\frac{3}{32}$-inch diameter and $\frac{3}{8}$-inch in length. I have also received taps which, when measured with a micrometer, were found to be very perceptibly smaller in diameter than the screws for which they were supposed to be intended.

Under these circumstances, the amateur will find it far better to obtain screws of Whitworth standard thread, and these may be had of Messrs. Davies & Timmens, Screw Manufacturers, 24, Charles Street, Hatton Garden, London, E.C. They are sold in quantities of a quarter gross and upwards. Nos. 134, 144, and 147 will be found the most useful sizes to keep in stock, and cost by the half-gross 1s. 6d., 1s. 8d., and 1s. 9d. respectively, while the postage for the gross and a half would be threepence. The prices per quarter gross are rather higher, and per whole gross rather lower in proportion to those given above.

Now as to taps. A screw-plate which will suit the above-quoted numbers of screws may be obtained from Messrs. Buck & Hickman, 280, Whitechapel Road, E., for 7s. This screw-plate is not sold as a Whitworth, but will prove to be the same thread. It has twenty-seven holes, and with it are sold a set of taps.

Plug taps the amateur must make for himself. They should not be like those usually sold, which are made from steel the same size as the tap required.

This plan will answer very well for taps of $\frac{1}{8}$ of an inch and upwards, but all taps under $\frac{1}{8}$ of an inch should be made from $\frac{1}{8}$-inch steel. The steel should be turned or filed down to the required size over the length that is to be screwed, but the thicker portion must graduate into the threaded part, and there must be no sharp angle. The other end of the steel should be filed to form a tang, as seen in Fig. 9, and

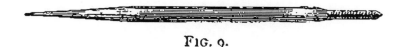

FIG. 9.

after the tap has been hardened and tempered, it should be fitted into a bradawl handle; it will then be found far more handy to use, and will last much longer than the little pieces of screwed steel wire to be purchased at the shops for 3d. and 6d. apiece.

Having at different times experienced considerable difficulty with the screws obtained from various sources for the reasons indicated above, I have been at some trouble to go fully into the matter for the benefit of my readers, and the above remarks embody the result of my investigations.

Now with regard to castings, the amateur may of course make his own patterns and get them cast at the nearest foundry. In such small castings the shrinkage may practically be ignored, amounting as it does to no more than $\frac{1}{8}$ of an inch in nine or ten inches. But in these days of cheap castings supplied ready to hand, I imagine there are few amateurs who will care to make their own patterns and take the risk of working up castings that have possibly been cast at a foundry where the workmen are unaccustomed to this class of work. For this reason I have not included a chapter on pattern-making in this volume; and indeed any amateur who feels a longing to make his own engine from the very beginning will learn more in half an hour from the inspection of a few patterns which will be shown him by any good-natured foreman of a foundry, than he would gather from a whole bookful of descriptive writing.

Supposing then that the castings are to be purchased from some one of the numerous firms who advertise these goods, the following points should be especially considered in making a selection. The castings should be even and free from holes, and should not present too rough a surface; also all the parts should be of such size and weight that when bored, turned, or filed down, as the case may be, they will still be thoroughly substantial, both in appearance and in fact. As amateurs (especially those living in the country) often find a difficulty in obtaining such things, I have written to several dealers, and am able to append the following particulars of the castings they supply. It must, however, be remembered that these particulars, except where otherwise stated, have been given me by the makers, and I cannot therefore guarantee their correctness further than by naming the sources from which I obtained the information. I subjoin the price asked by each maker for a certain size of engine, in order that my readers may have some idea of the various scales of charges.

Several of the engines described in this volume were made from castings supplied by Messrs. Lucas & Davies, of 21, Charles Street, Hatton Garden, who are themselves practical model-makers, and they will have much pleasure in showing to any amateur who will call upon them engines made up from sets of castings precisely similar to those they sell, so that the purchaser can judge of the appearance of the finished model before buying the castings. Messrs. Lucas & Davies charge for a set of horizontal slide-valve engine castings, one-inch bore and two-inch stroke, 4s., the bed-plate and columns for this set being 2s. 6d. extra.

Mr. R. A. Lee, of 76, High Holborn, supplies castings for models of all descriptions, from those for a single-action oscillating cylinder engine at 2s. the set, to those for a Great Northern express locomotive engine at £7 10s. All the castings supplied by this maker which require to be turned, are fitted with tenon pieces, so that they may be easily held in a grip-chuck for both turning and boring. Mr. Lee's catalogue, price 6d., will prove very useful to any of my readers who contemplate taking up the work in question. They will find that he charges for a set of horizontal slide-valve engine castings :—

Cylinder, one-inch bore, two-inch stroke—
    Best quality . . . . 6s.   set
    Common . . . . 4s. 6d. „
Cylinder, two-inch bore, four-inch stroke—
    Best quality . . . . 27s.   „

Mr. A. A. Dorrington, West Gorton, Manchester, has sent me his price list and some designs. He charges for castings of an engine—

Two-inch bore, three and a half-inch stroke—
                                  12s. 6d. set
Three-inch bore, four-inch stroke   28s. „
Working drawings, 2s. and 5s. respectively.

Mr. A. Wood, 15, Dalley Street, Lower Broughton, Manchester, supplies castings with tenon pieces to those parts requiring turning at the following rates:—

    One-inch bore, two-inch stroke . 3s. 10d. set
    One and a quarter-inch bore, two and a half-inch stroke . . . . . 7s. 6d. set

Mr. Wood's prices for boring cylinders and such work appear to be exceptionally low.

Mr. J. Tomlin, 28, Highfield Terrace, Barnsley, supplies castings and forgings of half-horse power, horizontal and launch engines.

    Cylinder, two and a half-inch bore, three and a half-inch stroke . . . . 12s. 6d. set

Copper tube and sheet and copper rivets, are supplied by Messrs. Stanton Bros., metal merchants and workers, of 73, Shoe Lane, London, E.C.

Introduction.

In case any of my readers should wish to obtain sets of castings exactly similar to those described, I may here mention that the castings for the engine described in Chap. II. were obtained from Mr. R. A. Lee, and the various sets mentioned in Chaps. VII. to IX., together with the separate castings described in Chapter XI., were supplied by Messrs. Lucas & Davies, of Charles Street, Hatton Garden.

CHAPTER II.

SINGLE-ACTION OSCILLATING CYLINDER ENGINE.

A DESCRIPTION of the fitting up of a model single-action oscillating cylinder engine is placed first in the series of models described in this volume, because, owing to its simplicity and to the fact that it is not necessary to bore the cylinder, few tools are required, and the beginner is more likely to succeed in its construction; while, at the same time, the cost of the requisite materials is so trifling that it is not a matter of great importance should any part be spoilt and have to be replaced.

Thus, although the single-action model has no useful archetype, it is itself useful as a stepping-stone to the modelling and comprehension of more complicated engines, especially those belonging to the oscillating class; and it may not be without interest to note that James Watt himself constructed at least one model of an oscillating engine, a drawing of which is given in Fig. 10.

The castings, etc., required for the model in ques-

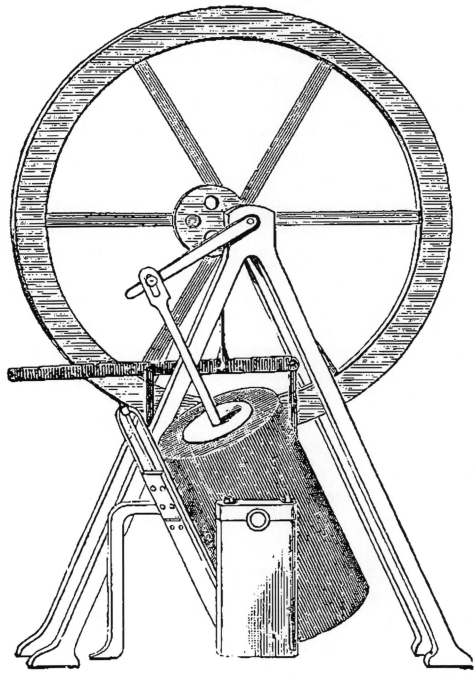

FIG. 10.

tion, together with a piece of brass from which to form the bed-plate, cost about two shillings. The castings consist of the fly-wheel, a pair of standards, the crank, steam-block, cylinder bottom and top, and piston; and a small piece of brass tubing is supplied with the castings for the body of the cylinder.

The cylinder being the most important part, we will attack it first.

Such a cylinder as the one under consideration— viz., ½-inch bore, and 1-inch stroke single action— will not require boring, but the interior must be smoothed out. In order to accomplish this, turn down a piece of soft wood about 5 inches in length to a slightly taper form at one end, in such a manner that the smallest part of the tapering end will pass easily through the small piece of tube which is to form the cylinder. From this end the piece of wood should increase in size, till it is just so large that it would require hammering to drive it through the cylinder, and from this point to the other end it should be cylindrical.

This piece of wood being turned, it is to be mounted between the lathe centres, with the piece of cylinder tubing upon the small end of it. Now coat the wood with a little fine emery and oil, and holding the bit of tube between the fingers, set the lathe in motion, and grind the tube out smooth. Before taking it out of the lathe, it may be forced up to that end of the wood which has not been used in the grinding process. This part will now be just large enough to hold it securely, while the ends are turned off true, and the outside polished with a piece of fine emery paper.

If the lathe be a good one, and the owner should not care to use it for grinding purposes, this part of the process may easily be carried out by hand.

The next part to be undertaken is the bottom of the cylinder. This piece must be held in a grip-chuck by the tenon attached to the under side of the casting, and the interior of the piece is to be turned out until the cylinder-tube will just slide stiffly into it, the bottom being turned out square, and the top edge neatly turned off either square or to a bevel from the outside, according to fancy. Now take the cylinder bottom out of the chuck, and placing it in the vice file off the tenon, leaving the bottom square and smooth. File up the outside all round, leaving it as smooth and truly round as possible; and file also the bearing-face, that is, the circular flat piece attached to one side of the casting, till it is smooth and level and as square as possible to the top, and consequently parallel with the inside. The centre of this face should now be found, and a circle struck from it as near the edge as can be managed; and then using this circle as a guide, the edge must be filed up round; or if the piece can be conveniently chucked for the purpose, it may be again mounted in the lathe, properly centred (taking care that the bearing-face is exactly at right angles to the mandrel), and the face and edge turned up. If this is properly done, this important part will require no further finishing.

The piston, a small circular casting with a shallow groove running round it, is now to be turned up. The turning of the lower face and edge can be done while the piece is held in the grip-chuck by its tenon, a flat groove being turned in the edge to take the packing; and while still in the lathe, it should be drilled for tapping with the screw for the piston-rod. The piston is now to be reversed, the tenon cut off, and the upper side turned. The centre hole may then be tapped, and the piece placed on one side for the present.

The top of the cylinder is formed from a casting very similar to that used for the piston, but rather larger. It is to be turned up in the same manner as the piston, but the edge, instead of being grooved, will be turned down on the lower side, leaving a flange on the upper side; and the edge, where turned down smaller, should be left very slightly conical, and of such a diameter that it will be a stiff fit in the cylinder tube. The top thus made forms a sort of plug in the upper opening of the cylinder; but it might also be turned out hollow, and fitted on to the cylinder as a cap. In either case the top must be drilled through while in the lathe. The tenon, instead of being cut off, may be turned up as shown in Fig.

12. This will give the cylinder a more finished appearance.

The steam-block (Fig. 11) and the crank have only at present to be filed up, and left as neat in shape and as smooth as possible.

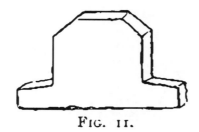

Fig. 11.

The fly-wheel must be centred, drilled, and mounted on a temporary shaft for turning. This will be found better in most cases than turning it on its own shaft, there being in the latter mode considerable risk of breaking the shaft, unless this latter is made so stout as to look clumsy in the finished model. The fly-wheel need only be turned upon its edge, the sides being made smooth with a file.

The standards and the bed-plate may now be filed up, the openings in the standards being worked out smooth with round and half-round files. The bed-plate must be squared up with the aid of the steel square.

Some screws and nuts will be required in the fitting together of the several parts, and these, if not purchased, should now be made. The nuts may be made of any piece of stout brass about one-tenth of an inch thick, as, for instance, an old clock-plate. From some such piece cut off with a metal saw, or otherwise, a strip 2½ inches long by ⅜ of an inch wide; mark this strip off into square pieces, and file about half-way through at each mark. Centre-punch each square, and drill through with a drill suited to the tap which is to be used, which latter should be about one-tenth of an inch in diameter. Now tap each hole, when you will find yourself supplied with a row of nuts which only require to be broken off from the strip and filed up at the edges. If preferred, the corners also may be filed off, and the nuts thus made octagonal.

For the screws, take a piece of brass wire about No. 6, New Standard Gauge, and file down half an inch at one end, till it is the right size to take the same gauge of screw-thread as that used for the nuts. Cut the thread with the screw-plate, and cut off the screw with a head about one-eighth of an inch deep. File the head up flat in the vice, and notch with a knife-file or saw; or if preferred, rather larger wire may be used in the first instance, and the heads filed up either square or octagonal, in which case they will not have to be notched. Of course, if a proper chuck is available, it will be quicker to turn down the wire for the screws in the lathe.

Now with the square mark on the bed-plate two lines exactly parallel, the first about a quarter of an inch from the edge of the bed-plate, the second about one inch distant from the first. Drill through the two feet of each standard with a drill a trifle larger than the screws you are using, so that the holes when drilled will allow the screws to pass through them. Place one of the standards on the bed-plate with the outside of its feet against the first line, and mark the bed-plate through the drilled holes. The second standard should now be placed with its feet just the further side of the second line; and again mark. This should leave about three-quarters of an inch clear space between the standards. It is as well to so mark the standards and bed-plate, that the standards may, in fitting up, be again placed without difficulty in their original position.

To mark the bed-plate it is a good plan to make a small centre-punch of a piece of steel wire the same size as the screws used, and about half an inch in length. This should be filed at the end to rather an obtuse but at the same time sharp point, and hardened; it is then to be placed in the hole in the standard or other piece, the position of which we desire to mark on the bedplate. It will stand upright without being held, thus leaving the left hand free to support the standard, and a sharp blow upon it with a light hammer will centre-punch the bed-plate ready for drilling.

The four holes for the standards may now be drilled in the bed-plate. These must be the same size as those already drilled in the standards; and the standards may now be bolted to the bed-plate, and tested with a square. They will probably be found to stand considerably out of the perpendicular, in which case they must be taken off the bed-plate, and the feet must be filed and fitted until the standards, when bolted to the bed-plate, stand perfectly perpendicular to it, and parallel with each other.

A centre mark is now to be made on each side in the centre of the solid top piece of each standard, and the standards being bolted to the bed-plate, and a piece of wood placed between them to prevent their giving under the pressure, the hole for the fly-wheel shaft may be bored in the lathe through both standards at one operation; this hole may be from an eighth to a tenth of an inch in diameter.

The standards can however, if preferred, be bored for the fly-wheel shaft before their position is marked upon the bed-plate. In this case the shaft also should be turned and finished before the position of the standards is determined, so that it can be placed in its bearings while marking the bed-plate through the feet of the standards. This is necessary in order to ensure the coincidence of the two bearings. The holes in the bed-plate may also take the form of short slots, so as to allow a slight adjustment to be made in the position of the standards.

The shaft may be rather larger than the bearings bored for it in the standards, it being turned down at the two ends to the necessary extent, thus leaving

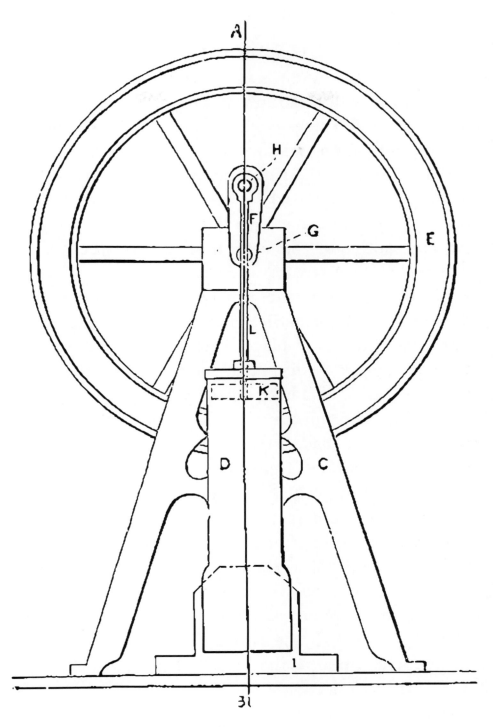

Fig. 12.

a shoulder to work against the inner side of each standard; or the shaft may be of the same diameter throughout, with a shallow and narrow groove turned in it where it passes through one of the standards, and in this case a small set screw must be fitted into the bearing from the side or top, and after the shaft is in position, this set screw is screwed in till its point just enters the groove in the shaft, thus preventing any motion of the latter from side to side.

The fly-wheel must fit the shaft perfectly. A keyway must be filed in the fly-wheel with a small square file, and a corresponding flat filed upon the shaft. After the wheel is in place, a small and very slightly tapered key or wedge must be driven into the space thus provided. During this operation the shaft should be held in the vice in an upright position, with pieces of lead between it and the jaws of the vice to preserve it from injury, and the fly-wheel being placed in position, the key is gently tapped in with a light hammer.

Before proceeding further, it will be found best to make a full-sized working drawing of the engine, as shown in Fig. 12, where A B is the perpendicular centre line, C one of the standards, D the cylinder, E the fly-wheel, F the crank, G the fly-wheel shaft, H the crank-pin, I the steam-block, K the piston, and L the piston-rod.

This drawing is made from an engine constructed from castings supplied by Mr. Lee, and it will be seen that the cylinder might with advantage have been cut shorter, the cylinder as sent out being adapted for a 1½-inch stroke, while the crank only allows a 1-inch stroke.

Fig. 13 shows the same engine as seen from above.

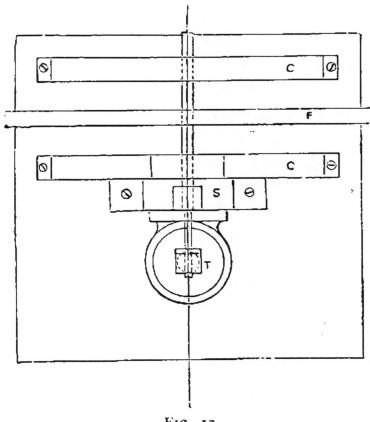

FIG. 13.

C C are the standards, F is the fly-wheel, S the steam-block, and T the top of cylinder.

Now, having our working drawing ready, we will proceed with the fitting. The bearing surface of the cylinder, if not turned up in a lathe, must now be finished off as level as possible. For this it should be filed up with a fine file, and then rubbed with water on a good flat stone. It is useful to have a quick cutting stone, and to keep it for this sort of work. One face of the steam-block must also be finished in the same way, and the holes for the screws by which the latter is to be bolted down to the bed-plate may also be bored at this stage of the proceedings.

A line should now be marked down the centre of the bearing surface of the cylinder, and another through the centre of the face of the steam-block, as seen at A B, Figs. 14 and 15. Upon the former line punch two centre-marks on the cylinder face, in about the positions marked D and E in Fig. 15; also centre-punch the crank for the crank-pin and fly-wheel shaft.

Now draw the diagram as shown in Fig. 16, as follows: at right angles to the straight line A B draw C D, equal in length to the distance between centre-marks on crank, and draw D E equal to distance between crank-pin H (Fig. 12) and centre-mark D (Fig. 15), when the engine is in the middle of its stroke. (This may be easily found by adding dotted lines to Fig. 12, showing positions of cylinder, piston, and crank at half-stroke, and of centre-mark D.) Continue D E to F, equal to distance between centre-marks D E (Fig. 15).

Now taking the steam-block (Fig. 14), centre-punch

on the flattened side the mark F to correspond with the mark D on cylinder face, and with the radius F K equal to D E (Fig. 15) describe the arc H K. Turning to the diagram Fig. 16, take the distance F H with

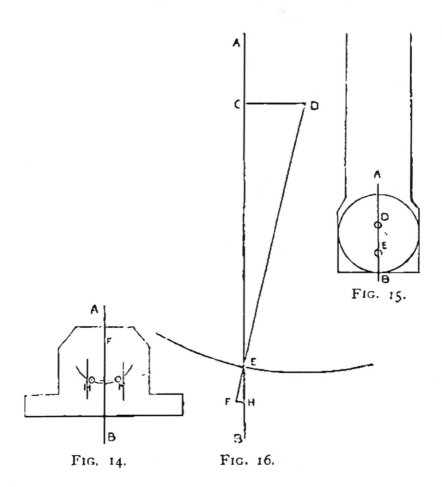

FIG. 14. FIG. 15. FIG. 16.

compass, and mark off this distance on Fig. 14 on each side of the centre-line A B. The centre-marks for the steam-ways must be made upon the arc H K, and just within the distance marked off from the centre-line, so that the steam-ways themselves, when drilled, may be just within the mark, as seen at H and K (Fig. 14). The steam-ways are to be drilled half-way through the steam-block with a $\frac{1}{18}$-inch drill; a rather larger drill is then to be taken, and with it a hole is to be drilled from each side of the steam-block into the $\frac{1}{18}$-inch steam-way just mentioned. This will be made quite clear by a glance at Figs. 17 and 18, Fig. 18 being a section of the steam-block on the line A B of Fig. 17, and H and K the steam-ways.

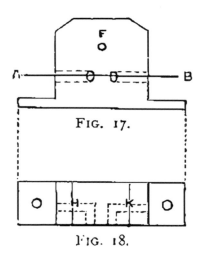

Fig. 17.

Fig. 18.

Bore the holes D and E in cylinder-face with a $\frac{1}{18}$-inch drill, as marked in Fig. 15, and tap a screw in D. Take a piece of brass wire one inch long, and screw about $\frac{1}{4}$ inch at one end to fit D, and $\frac{1}{2}$ inch at the other end; this is to be screwed into the hole D in the cylinder-face, and forms the pivot upon which the cylinder will oscillate. Take a piece of brass, rather under half an inch in diameter, and after boring it to take the same thread as that with which the hole D (Fig. 15) has been screwed, turn it down, partly cut off about one-eighth of an inch, mill the edge, finish cutting it off with a parting-tool, and screw the hole in the centre with the same tap as that used for D. Now take a piece of wire a trifle larger than the piece used for the cylinder pivot, and round it twist a piece of cold-drawn brass wire to form a spiral spring half an inch in length.

Bore the hole F in the steam-block (Fig. 17) a trifle larger than D (Fig. 15), so that the cylinder pivot will pass through it easily, yet without having room for play.

Tap the ends of the steam-ways at the sides of the steam-block for steam and exhaust pipes respectively.

Give the cylinder-face and surface of steam-block a final rub with a little oil upon a good flat oil-stone to take off the burrs at the edges of the drilled holes, etc. Probably no great difficulty will be experienced in getting the surfaces true and flat. To ascertain whether you have succeeded in obtaining a good fit, smear one of the surfaces with a little red lead, and then press the other against it.

Screw the pivot into its place in the bearing-face of the cylinder, push it through the hole bored for it through the steam-block, pass the spiral spring over the end of the wire projecting on the other side of the steam-block, and screw on the small milled nut. It will be seen that by means of this nut the tension of the spiral spring can be altered, and the surfaces of the cylinder-face and steam-block brought together with more or less pressure as required.

The crank casting may now be bored and tapped, and the shaft cut off to project $\frac{3}{8}$ of an inch beyond its bearings, and one end screwed to take the crank. The crank-pin is formed of a small piece of steel wire, $\frac{1}{2}$ or $\frac{5}{8}$ of an inch long, one end being screwed to fit the casting. Bore a piece of brass measuring about $\frac{3}{8}$ by $\frac{1}{4}$ of an inch with a hole a trifle larger than crank-pin. File up this piece of brass square, or, if preferred, it may be turned off a piece of rod in the same manner as the tension nut was formed. Drill a hole in the edge, and tap this to take the top of the piston-rod, which must also be screwed to fit, after it has been cut off to the right length; this may be ascertained by a reference to the working drawing.

The steam-block may now be placed in position upon the bed-plate, with the crank upon the shaft, and the head of the piston-rod upon the crank-pin; and taking care that the centre upon which the piston works is directly under and parallel with the shaft of the fly-wheel, mark the position for the holes in the bed-plate which are to receive the screws by which the steam-block is to be held in place. It will probably be found that the steam-block comes partly between the legs of the front standard.

The holes in the bed-plate are now to be drilled, and the steam-block bolted down to it. It may be found necessary to make the heads of the bolts which secure the steam-block very shallow, so that they may not interfere with the screwing in of the steam and exhaust pipes. The piston must be packed with a little tow, and oiled; and now, with a little oil upon the shaft of the fly-wheel and on the cylinder bearings, our engine stands complete and in working order.

CHAPTER III.

SINGLE-ACTION OSCILLATING CYLINDER LOCOMOTIVE.

THIS, the most simple form of steam locomotive, must not be despised by the amateur worker; for although it is not like the slide-valve locomotive, an actual model of a full-sized engine, its construction will give some good practice in fitting, while it may be finished in a far shorter time than the more complicated slide-valve locomotive.

The castings for the engine now in question cost 6s. 6d., and the following parts are comprised in it:—

The bed-plate or frame, seen in Fig. 23,
Piece of brass tube to form the body of the boiler,
Two cast boiler ends,
Piece of brass tube for the chimney,
Casting for top of chimney,
Steam-dome,
Two number-plates,
Two buffers (cast in one piece),
Two driving-wheels,
Two small wheels,
Two cylinder blocks (cast in one piece, Fig. 19),

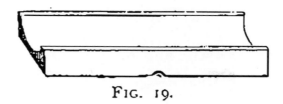

Fig. 19.

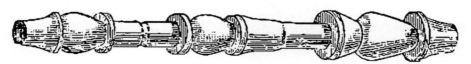

Fig. 20.

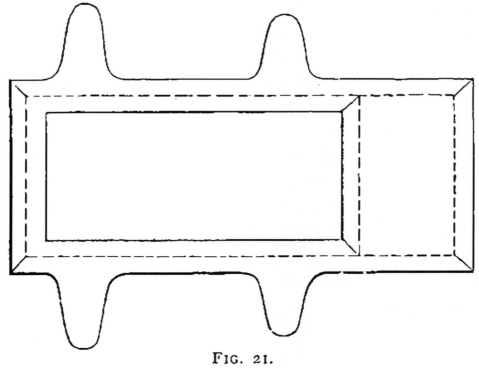

Fig. 21.

and a casting (Fig. 20), from which may be formed the steam-tap, whistle, and gauge-tap.

An engine of this class can very well be made without any castings except for the wheels; in that case the bed-plate is cut out of stout brass and turned down as shown by the dotted lines in Fig. 21. Steam-blocks are soldered on and connected by a steam-pipe, into the centre of which the pipe from the boiler is soldered. A prettier model, however, and one that will prove in every way more satisfactory, will be turned out if the proper castings are purchased.

The first thing to be done is of course the filing up of the separate parts, and we will commence with the bed-plate. This should be filed up bright all over, the grooves in the foot-plate being left as cast, or filed out smooth with a triangular file, according to taste. The steam-blocks on each side must be filed smooth, flat, and perfectly square with the top of the bed-plate. The four wheels must be turned up; these present no difficulty, only it must be remembered that each pair must be of exactly the same outside diameter, otherwise the engine will not run straight without rails. When the wheels have been turned, the bed-plate may be bored for the axles. The position of the bearings in the lugs cast for them on the underside of the bed-plate may easily be found in the following manner:—

Make a full-sized drawing of one side of the bed-plate, and (parallel with the top of the bed-plate) draw a line rather more than the diameter of the small wheels below the bed-plate. Now draw two circles to represent the wheels, just touching with their circumference the lower parallel line, and the centres of these circles will mark the positions for the axle-bearings. The axles will be made of steel rather over $\frac{1}{8}$ of an inch in diameter, and the wheels may be either keyed on or screwed, the latter method being the more convenient, as the wheels can then be easily taken off when required. The carriage should at this point be put together and tested by running it on a smooth and level surface, to see if it runs straight.

The boiler may now be taken in hand. The top must be soldered on to the funnel-tube, and this part may then be put into the lathe and polished with emery-paper; the steam-dome must be turned up and the under side filed to fit the curve of the boiler. Two holes must be drilled exactly opposite to each other at one end of the boiler-tube, to take the funnel which may be at once soldered in. If the steam-pipe is to be inserted into the dome, a hole $\frac{1}{4}$ of an inch in diameter must be made in the boiler, and another $\frac{3}{16}$ of an inch in diameter in the centre of the dome to receive the steam-pipe. These holes must of course be made before the dome is soldered on; and when this is done, care must be taken to have the dome in a line with the funnel. Between the dome and the end of the boiler-tube let there be soldered on to the inside of the tube a piece of brass as thick as the tube itself or somewhat thicker, and half an inch or so square, then bore and tap the part thus lined for a quarter-inch screw. This is the man-hole, and although the size given here is smaller than usual, the engine will have a neater appearance; the small man-hole is more easily kept steam-tight, and with a short length of indiarubber tubing to act as a siphon, there need be no difficulty about filling the boiler.

The ends of the boiler must next be filed up and soldered on. A little solder should be run round the flange of each end-piece and round each end of the boiler-tube itself. Now stand the boiler-tube in position upon the end to be soldered, and a soldering-iron or blow-pipe flame run round the joint will unite the two parts.

The cylinder-tubes will next be ground out smooth, as described in Chap. II. Two pieces of sheet brass must be filed up to fit into the cylinder ends, and one soldered into one end of each.

The steam-blocks (Fig. 19) must then be sawn apart, and the groove in each block filed out to fit the cylinder-tubes, which must then be soldered to the blocks. A hole $\frac{1}{16}$ of an inch diameter is to be drilled through the steam-block into the cylinder close to its lower end; and about $\frac{3}{16}$ of an inch above it, a pin made of steel wire rather over $\frac{1}{16}$ of an inch

Single-Action Oscillating Cylinder Locomotive. 35

in diameter and $\frac{3}{4}$ of an inch in length, is to be screwed into the cylinder face. This pin must also be screwed at its free end to take a small nut.

The cylinder covers must be turned up, and should, when finished, fit stiffly over the tops of the cylinders, the hole for the piston rod being of course bored while the cover is in the lathe. The pistons should be turned up from a piece of brass rod, and bored and tapped for the piston-rods; the latter are made of steel, $\frac{1}{16}$ to $\frac{3}{32}$ of an inch in diameter and 2 inches in length. Heads for the piston-rods are turned up from $\frac{1}{4}$ inch brass rod to the shape shown,

a hole being drilled through each to take the driving-pins of the driving-wheels.

The driving-pins are made of pieces of steel wire the same size as the piston-rods, and about half an inch long. They must be screwed into the solid part of the driving-wheels, at a distance of half the piston-stroke from the centre of the wheels.

The steam-ways may now be made. It will be found that, underneath the rear end of the bed-plate casting, the metal is thick enough to allow of the steam-ways being bored in it. A hole $\frac{1}{16}$ of an inch in diameter is first bored from one steam-block half-way through, and a similar hole from the other steam-block joins it. These steam-ways should be bored as low down as they safely can. The hole for the steam-pipe is made underneath, if the engine is to be fitted up as shown in Fig. 24, or it may be made from the top, and the steam-pipe brought to it from the end of the boiler instead of coming from the dome at the top. A line must now be drawn through the centre of the steam-way and the driving-pin, with the driving-wheel in the position it would occupy with the piston at half-stroke. Upon this line should be drilled the hole for the cylinder-pin, of course at the same distance from the steam-way as the pin itself is from the steam-port in the cylinder. Grooves are cut in the steam-blocks for the exhaust steam, as shown at A in Fig. 22. To find the posi-

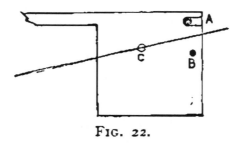

FIG. 22.

tion of the exhaust, draw a line running through the centres of cylinder-bearing and driving-wheel, and the exhaust will be the same distance above such a line as the steam-way is below it.

The cylinder face must be ground to the steam-block with a little fine emery and oil, the cylinder being worked backwards and forwards as it will ultimately work under steam, and the surfaces may then

Single-Action Oscillating Cylinder Locomotive. 37

be washed and finished off by working with a little putty powder in place of the emery. The cylinder faces are kept against the steam-blocks by means of spiral springs and nuts, as explained in the chapter on an oscillating cylinder engine. One of these springs may be seen in Fig. 23.

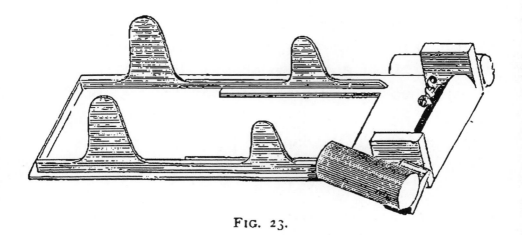

FIG. 23.

The engine may now be fitted together, the steam-pipe being soldered into the boiler, and into the hole made for it in the bed-plate. If a safety-valve is to be added to the boiler, a steam-tap may also be added; but if no safety-valve is put in, it is better to connect the boiler direct with the steam-ways.

The screw for the man-hole of the boiler should be made with a rounded head milled at the edge; a little hemp or a leather washer will make it steam-tight. The number-plates have each a flange by which they are to be soldered on.

In putting on the driving-wheels, be careful to see

that when the driving-pin of one wheel is at its highest point that of the other wheel is at its lowest. If the wheels are screwed on to the axles, it may be necessary to file a flat upon one of each pair, and drive in a wedge when the wheels are in the correct position. With only one wheel of each pair to screw on, the wheels will still be easily taken off, should this at any time be necessary.

A small slip of mahogany should be screwed on to the front of the engine-frame, and the buffers having been turned up and separated are fastened into two holes in the wood. Fig. 24 shows the finished engine.

Fig. 25 shows the lamp; this is made in the following manner. Make first a tin box about 2 by $1\frac{1}{2}$ by 1 inch,—for material a piece of an empty meat-tin will do as well as anything. Then a piece of $\frac{3}{8}$ inch brass tubing, $2\frac{3}{4}$ inches long, is bent as shown in the illustration, and a short piece half an inch long is soldered into it about an inch from the curved end. This branched tube is now soldered into the tin-box, and to the other end of the box is soldered a hook made of a piece of wire doubled and bent, by which the lamp may be hooked on to the foot-plate of the engine, while another piece of wire soldered to the front of the lamp rests upon the axle of the front wheels. The hole in the top of the lamp for filling may be closed either with a cork or a screw. A small pinhole should be made in the top of the lamp to allow air to enter as the spirit is burnt away.

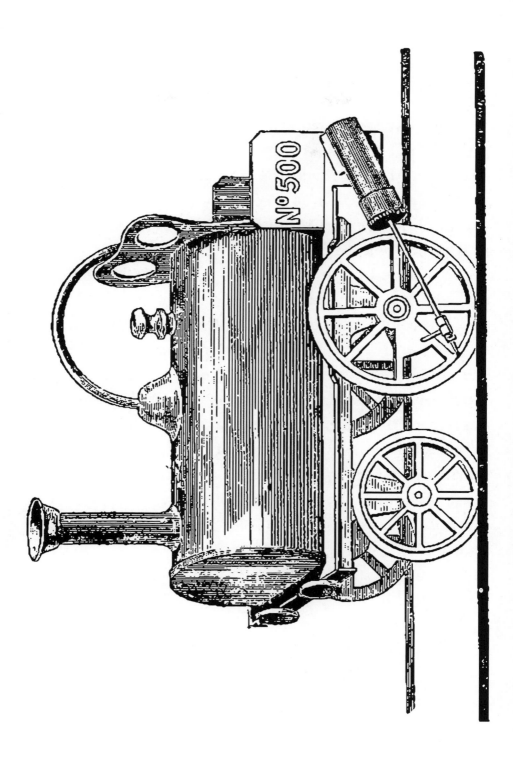

40 *Single-Action Oscillating Cylinder Locomotive.*

The casting for the taps and whistle (Fig. 20), is given here, because being supplied with the set the reader might otherwise be puzzled as to its use; but these adjuncts are not necessary to this engine, and the method of making them will be fully described in another chapter.

A double-action oscillating cylinder locomotive may be made by using the double-action cylinders described in the chapter dealing with the marine engine; in this case the cylinders may be placed on the outside of the frame, and kept up to the steam-block by means of spiral springs as in the engine just described, or they may be placed inside the frame and kept in place by means of set-screws as in the marine engine; in the latter case the driving axle must of course be provided with two cranks. The method of making a double crank is described in another place.

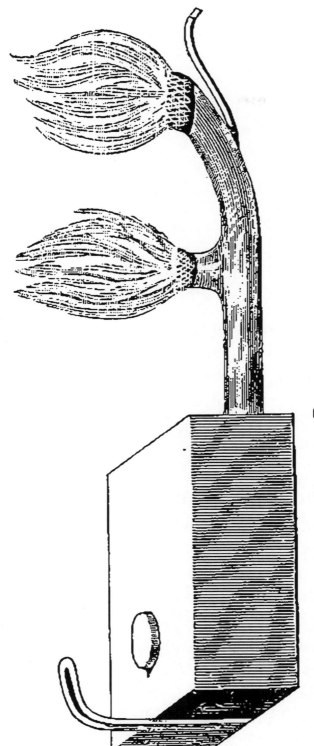

Fig. 25.

CHAPTER IV.

HORIZONTAL SLIDE-VALVE ENGINE.

THE CYLINDER.—CYLINDER COVERS.—STEAM-CHEST AND STUFFING-BOXES.

THIS engine will present considerably greater difficulties in construction than those described in previous chapters; indeed, it will tax to the utmost the amateur's capabilities, whether in turning, drilling, filing, or fitting. Still, with careful attention to the smoothness, flatness, squareness, and straightness of the working parts, all obstacles may be overcome, and the result will not only be a model of handsome appearance, but one which works upon exactly the same principles as the half and one-horse power engines in every-day use in many small workshops; and indeed the mechanism of the most powerful locomotive is but an adaptation with additions of the model we are now about to construct.

It is not advisable to take too small a size for the first model made; perhaps the most easily managed size is a model having a cylinder of one-inch bore and two-inch stroke. In a smaller size, the minute-

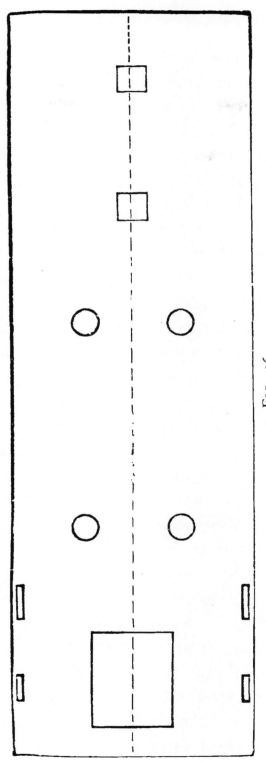

FIG. 26.

ness of the details and the delicacy of some of the parts might lead to difficulties; while, on the other hand, in the larger sizes we find that the greater extent of some of the surfaces requiring to be made level and steam-tight renders it less easy to obtain the needful accuracy in fitting and smoothness in working.

When purchasing the castings, the cylinder may be had ready bored at an additional cost of about 2*s.*; and I recommend all who are not possessed of a good lathe, and are not also very certain of their own powers, to get this important part of the work done for them. For the benefit, however, of those who prefer doing everything for themselves at all risks, I shall describe several methods of boring the cylinder, depending upon the apparatus available.

Fig. 26 shows the bed-plate, and Figs. 27 to 41 show most of the castings usually supplied in what is called a set; to those shown in the engravings must be added a fly-wheel, which requires no illustration. It may here be mentioned that Figs. 31, 32, and 33 are supplied in pairs, and that several of the smaller parts in these sets are often cast in one piece; thus, Fig. 34 is really a casting of four small pillars in one length, Fig. 29 will form the two guide-blocks, and Fig. 41 the glands for the stuffing-boxes of the cylinder and steam-chest respectively. Fig. 40 is sometimes omitted from the set; in this case it must be worked up from a forging, which any blacksmith

Fig. 27.

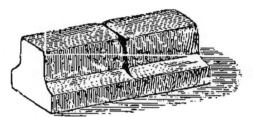

Fig. 29.

Fig. 28.

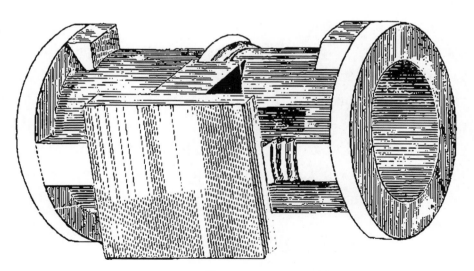

Fig. 30.

will turn out if supplied with a full-size drawing of what is required. The crank (Fig. 42) is worked up from a forging.

THE CYLINDER.—This is by far the most important part of the engine, for it is in the cylinder that the steam exerts its expansive power; or, in other words, it is here that the heat generated in the furnace is turned into work; and upon the accuracy with which the cylinder is bored and fitted, largely depends the effectiveness of the engine. Even in a model, where the highest degree of effectiveness is not a consideration, it will be impossible to get our engine to work evenly and well unless the most careful attention has been given to this part of the construction. Those of my readers who reside in the neighbourhood of London, and who, after procuring their castings, find themselves unable to get over the boring of the cylinder, to their satisfaction, will find Mr. Smelt, of 7, Goldsmith's Row, Gough Square, very attentive to their requirements. Mr. Smelt will undertake any machine work, either small or large, at a moderate price, and will, I think, be glad to render any assistance in his power to amateurs in difficulty. However, there is no insuperable difficulty in boring the cylinder at home.

The first step of course is to chuck the cylinder, and this may be done in two or three ways. First, it may be attached to the face-plate thus: take a circular piece of wood about a quarter of an inch thick

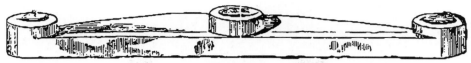

FIG. 31.

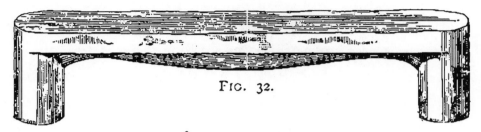

FIG. 32.

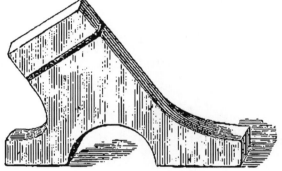

FIG. 33.

FIG. 34.

FIG. 35.

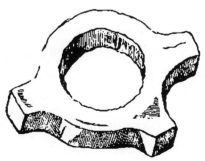

FIG. 36.

and of rather greater diameter than the cylinder end, including the flange. If this piece of wood is turned up, and a few concentric circles marked upon its face, it will facilitate the centring of the cylinder. Now place the piece of wood upon the face-plate, and then after filing one end of the cylinder as flat and square as possible, place it filed end downwards upon the wood. Next take three dogs, *i.e.*, pieces of iron or steel about two inches long and three-eighths to half an inch thick, with a hole in the middle of each to take a bolt. Place these with one end of each resting on the flange of the cylinder, and directly over three of the slots in the face-plate. Now place a small piece of wood about half an inch thick under the other end of each dog, and finally pass a bolt up through the slot in the face-plate and through the hole in each dog, and screw the nuts down so that the cylinder is firmly held in place, but do not yet tighten the nuts ready for action. Screw the face-plate with cylinder attached on to the lathe mandrel, and see whether the cylinder runs true. Probably it won't, so we first centre the back flange by holding a piece of chalk against the cylinder, as near the flange as possible, and giving a slight tap where required with a light hammer or mallet, till the back flange appears properly centred. We now test the front flange, and if the back flange, besides being truly centred, has been filed square and flat, the front will run true ; otherwise we must get it right by loosening slightly the

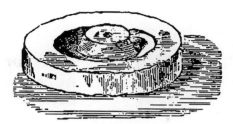

FIG. 37.

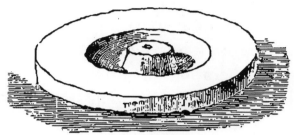

FIG. 38.

FIG. 39.

FIG. 40.

FIG. 41.

nuts as required, but only one at a time, and packing up whichever side requires it with thin pieces of metal, taking care in the meantime that the position of the cylinder upon the face-plate is left unaltered. The arrangement just described is shown by Fig. 43; and when it is completed and the cylinder satisfactorily centred, the nuts must be carefully tightened up, so that there may be no danger of the work shifting during the operation of boring.

Another method of chucking the cylinder, one which may be adopted by those who do not possess a face-plate, is by turning up a piece of beech or box-wood, and turning in it a recess into which the cylinder-flange will fit exactly. To prevent accidents, it is well to secure the cylinder in this recess by the same sort of dogs as those used in the method first described, ordinary wood screws being passed through the holes in the dogs into the wooden chuck; or three pieces of iron or wood having a hole at each end may be fastened across the flange of cylinder by screws, as shown in Fig. 44.

Yet another way of securing the cylinder is by means of angle-plates used with the face-plate; but this method is more useful for cylinders of larger size than the one now under consideration.

If the cylinder is to be bored with a boring bar the first two methods will have to be slightly modified, as thicker wood will have to be used; and a recess, the diameter of bore of cylinder when finished, must be

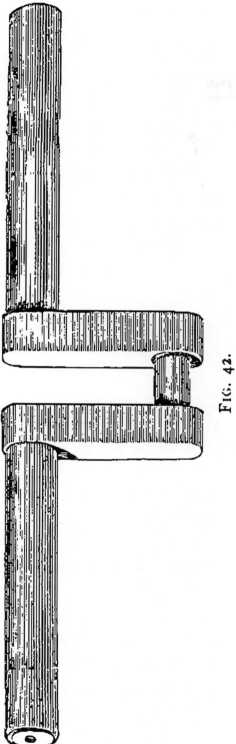

Fig. 42.

turned in it to a depth of three-quarters of an inch, in order that the boring tool may pass completely through the cylinder.

This last method will probably suit most amateurs, and I shall therefore describe a simple and easily made boring bar with which the boring may be done.

Another way I may mention, in passing, is to fix the cylinder to the saddle of lathe instead of to mandrel, and use a boring bar; but those who possess the necessary apparatus for this are not likely to require any instructions as to how it is to be used. I shall therefore confine myself to such methods as may be of use to amateurs less practised than we should suppose those to be who are thus fully equipped.

Whichever plan of chucking has been adopted, we will suppose the cylinder to be firmly fixed and properly centred, and the next operation will be the actual boring; and if the lathe is provided with a slide-rest, this may be done with an inside tool held in the slide-rest. Two cuts will be necessary. The slide-rest must be in good condition, and great care must be taken to set it parallel with the work, otherwise the cylinder will turn out taper and be useless until it has been rebored, supposing there is enough metal remaining for that operation.

However, it does not fall to the lot of every amateur model-engine maker to be the happy possessor of a good slide-rest; therefore such of my readers as are

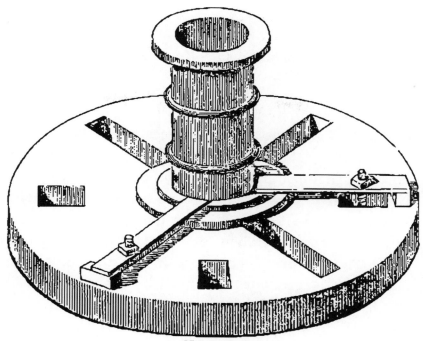

Fig. 43.

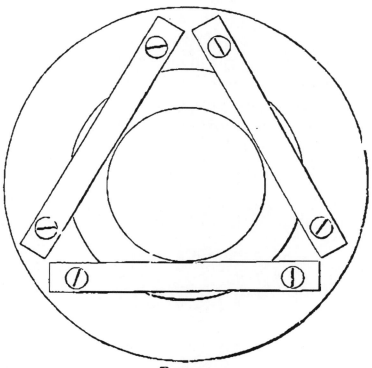

Fig. 44.

minus this useful adjunct to the lathe will be glad to hear that they may bore their cylinders in the following manner:—Take a piece of beech-wood about twice the length of the cylinder to be bored, and turn it down to the inside diameter of the cylinder casting. Now take two pieces of thin steel; a piece of an old saw will do very well. These may if necessary be softened, and must then be filed to a rectangular shape three-quarters of an inch in length; for about half of this length, one piece must be exactly one inch and the other a shade under one inch in breadth; for the other three-eighths of an inch, the width is to be gradually and very slightly narrowed, as at A (Fig. 45). A saw-cut is now to be made in the end of the piece of turned beech-wood, so that the pieces of steel will just fit tightly into it. When these are placed in position, first one and afterwards the other piece of steel (if only one piece of beech is used, but it is better to make use of a separate piece of wood for each cutter), a hole must be drilled through both wood and steel for the reception of an ordinary wood screw, the purpose of which is to prevent the steel from

FIG. 45A.   FIG. 45B.

shifting or the wood from splitting. This hole must of course be countersunk at one end to take the head of the screw (see B, Fig. 45). A slip of wood must be taken away with a chisel on each side, as shown in Fig. 46, to give clearance to the borings. The steel cutters should now be hardened, tempered, and pushed into their places, and a screw put through as already explained.

The mouth of the cylinder must now be turned up, so that the borer may start true; the borer is then placed in position, and the moving head-stock brought up to the back end of it. The lathe must be driven at a moderate speed, and the cutter sent through the cylinder, if possible, without stopping, at a uniform speed by means of the screw in the moving head-stock. After the first rough cut, the cutter is to be

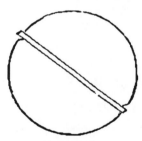

FIG. 46.

changed, and the boring finished by sending the full-sized cutter through in the same manner. The first cut may also be taken with a cutter filed up from a piece of square steel of the shape shown in Fig. 47. This cutter is to be driven through a hole bored in a piece of beech similar to that previously used.

Another method of boring the cylinder, similar to the last described, is with a D-bit; sections of this

FIG. 47.

tool are shown in Figs. 6 and 7. A bit of the proper diameter may be obtained from any engineering shop, made to order, or from Messrs. Buck & Hickman, Whitechapel Road, the cost being about 1s. for every eighth of an inch diameter. The D-bit might also be made at home, as follows:—A piece of cast-steel bar of the proper size is softened and one end slightly bent over, and then filed so that its surface may be again brought to a right angle with the length of the bar; this end is centre-punched close to the edge which, owing to the bending over of the end, projects beyond the rest of the bar. At the other end the centre-mark is punched in the centre of the bar. Now drill for the lathe-centres, and mount the bar in the lathe; if the bar has been properly bent over and centre-punched, one end will revolve upon its centre; but at a short distance from the other end,—where the bend is, in fact,—the whole thickness of the bar will be just on one side of a line joining the lathe-centres, and as a consequence will revolve round that line; and if the bar be now turned down with the slide-rest, the part near the end will be semi-circular in section, and will, when the bent portion is cut off, form the cutting edge of the bit. The bit will, of course, require to be hardened and tempered before it is used. The amateur who is unacquainted with the D-bit should purchase one in the first place, and will thenceforth, with the aid of the above description, find no difficulty in making his own.

A mandrel, upon which the cylinder is to be chucked while the two flanges are turned up, must now be turned from a piece of box or beech-wood, and the cylinder-ends finished upon it, unless the boring has been done with the slide-rest, for in this case one flange can be turned up at the time of boring, and the cylinder may then be reversed and chucked as before, and the other flange turned up.

The cylinder-face must be filed down smooth, and the cylinder then stood up on end on a surface-plate, or lacking that implement, upon the bed of the lathe, and the face tested with a square. File and scraper must be used until the face is perfectly square with the ends of the cylinder. A little red-lead and oil should now be smeared over the surface-plate or other flat surface used as a test, and the surface of the cylinder-face tried upon it. The red-lead will show which are the high parts, and these must be filed or scraped down until the surface is perfectly flat. While thus getting the cylinder-face flat, we must not, however, forget to test it occasionally with the square, otherwise by the time it is flat we may find that it has been filed and scraped out of square. Finally, to take out any scratches left by the file, the cylinder-face may be lightly rubbed on a flat stone. The lugs or feet of the cylinder must be filed up square with the face.

STEAM-PORTS.—The next operation to be undertaken is the marking out of the steam and exhaust-ports. The usual practice in large engines is to make the area of the steam-ports ·057 (or about one-twentieth of the area of the cylinder), and the length of the ports is taken as seven-tenths of the diameter of the cylinder. In this small engine the nearest approach to theoretical correctness is to make the steam-ports $\frac{3}{32}$ by $\frac{3}{8}$ of an inch.

To mark out the steam-ports it is best to use a scribing-block and surface-plate, but failing these tools a pair of dividers or calipers may be used. Before commencing to set out the ports, and especially if dividers or calipers are used, make sure that the cylinder-face is itself square. Now stand the cylinder on end on the surface-plate, and set the point of the scriber as near the centre as possible; mark the cylinder-face very lightly with the point of the scriber, and then, without moving the latter, reverse the cylinder, standing it on its other end, and again mark the face. The centre of the face will of course be a point midway between the two marks. Now place the cylinder horizontally on its feet, which should have been previously filed up square with its face, and again set the scriber-point as near the centre as possible and mark; reverse the cylinder, placing small blocks of wood under the face to keep it steady and square with the surface-plate, mark once more with the scriber-point, and rule centre-line between the two points marked. The first or vertical centre-line should, when drawn, be continued across the face of the cylinder and the boss for exhaust-pipe, while the horizontal centre-line must be continued along the metal left for the steam-ways and across the flange of the cylinder at each end.

Should the feet of the cylinder stand out beyond the flanges, or the flanges themselves be partly filed away during the process of filing the feet square with the cylinder-face, allowance for such difference must of course be made when marking the horizontal centre-line. This precaution will not be necessary if the face is marked out with the dividers, as the horizontal centre-line will then be marked from the edge of the cylinder-face, the vertical centre-line being taken from the flanges of the cylinder. Now with the scriber or by means of the dividers draw six lines, three on each side of the vertical centre-line, the first pair being each a full $\frac{1}{16}$ of an inch from centre-line, and the second and third pair $\frac{3}{32}$ of an inch from the first and second pair respectively. Now mark two lines, each $\frac{3}{16}$ of an inch from the horizontal centre-line; the ports will now be all marked out, and the cylinder-face will exhibit the

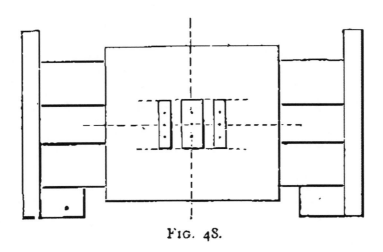

Fig. 48.

appearance shown in Fig. 48.* Now with a punch make three centre-marks in each of the spaces marked out for the ports, one mark being made in each case in the centre of the space marked out, and the other two each $\frac{1}{8}$ of an inch from this centre, one on each side. These centre-marks are also shown in the figure to which we have just referred. We next centre-mark for the steam-ways on each flange. These centre-marks must of course be made upon the continuation of the horizontal centre-line already

* Owing to a valve casting belonging to another set having been accidentally sent by the manufacturers, and Figs. 48 and 49 drawn therefrom, these figures do not in all their measurements answer to the description given in the text.

marked, and must be made on about the middle line of the thickness of the metal in which the steam-ways are to be bored, but if anything rather nearer to the inner than the outer side.

With a $\frac{1}{8}$-inch drill the steam-ways should be drilled down to within $\frac{1}{4}$ of an inch of the centre of cylinder-face, the drill being inclined slightly outwards, so that the steam-ways may run towards the surface of the cylinder-face. With the same drill make the three holes for the exhaust-port, as already marked. These holes should be rather under $\frac{1}{4}$ of an inch in depth; and next, using still the same $\frac{1}{8}$-inch drill, or one slightly larger, make the exhaust-way through the boss into the three holes just drilled. With a $\frac{3}{32}$-inch drill make three holes in each steam-port as marked. These holes should be drilled outwards towards the ends of the cylinder, to avoid any risk of their running into the exhaust-port; the two

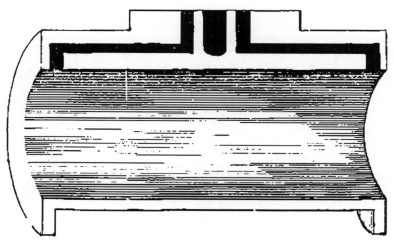

FIG. 49.

sets of holes must of course be drilled down into their respective steam-ways. The ports and steam-ways are shown in section in Fig. 49.

The ports may be left as they now are; but they will look more workmanlike, and the engine will also work better, if they are squared out to the lines marked with a small chisel. If it is intended to square out the ports, it may be advantageous to drill two steam-ways, each $\frac{3}{32}$ of an inch in diameter (instead of the one $\frac{1}{8}$-inch) at each end of the cylinder, for the two $\frac{3}{32}$-inch ways will have a rather larger total area than the square-cut port, while the $\frac{1}{8}$-inch steam-way will have an area rather less than the square ports.

The side of the cylinder must be chipped down a little, so as to make a clear way between the end of the steam-ways and the interior of the cylinder when the cylinder-covers are on; or a way may be made with a drill, and the end of the steam-way showing in the cylinder-flange may be stopped up with a small piece of brass wire.

CYLINDER-COVERS.—The bottom cover should be turned before the top one, since it is not so necessary that the former should fit the cylinder exactly as it is that the latter should be a perfect fit. If a chucking-tenon has been cast on the cylinder-bottom, the chucking will of course present no difficulty. It may be held in a self-centreing or grip-chuck, or a chuck may be made once for all by boring out a gun-metal chuck to the proper size and tapping it with a screw, which need not be a very fine one; the tenons may then be all screwed to fit this chuck, so that if the said tenons have been properly cast, most of the turning work may be done with the one chuck. If there is no tenon piece the chucking of the cylinder-bottom will be a less easy matter; but here, as is often the

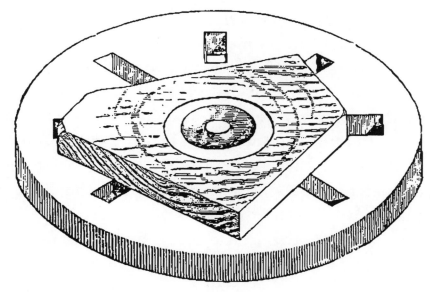

FIG. 50.

case, the simplest method will be found the best. Care must, however, be taken to perfectly fit the cover to the chuck before beginning, for there are few things more annoying than for a piece to work loose in the chuck just as it is half turned, and to keep coming out during the remainder of the operation.

To chuck the cover, take a piece of suitable wood —beech is good, boxwood preferable; but when nothing better was at hand, the writer has often used deal, and made it serve the purpose well enough; in fact, Fig. 50 is an exact representation of a cylinder-cover at the present moment before the writer's eyes, chucked in an odd piece of $\frac{3}{4}$-inch deal board, screwed to the face-place by two ordinary wood screws passed through holes in the face-plate and into the wood. However, to return to the chuck we are about to make,—the piece of wood chosen need not be over $\frac{3}{4}$ of an inch thick and $2\frac{1}{4}$ inches in diameter; it should be screwed to the face-plate with two or three ordinary screws, and if the face-plate is not already provided with the necessary holes, it will be found very convenient to drill half a dozen at different distances from its centre, these holes being about $\frac{3}{16}$ of an inch in diameter, and slightly countersunk at the back of the face-plate to take the heads of the screws.

The wood for the chuck being firmly attached to the face-plate and the latter screwed on to the lathe-mandrel, a recess of $\frac{3}{16}$-inch depth should be turned in the centre, and this recess must be gradually enlarged until it is nearly of the same diameter as the cylinder-cover. The edges may be very slightly undercut. When the cover will all but go into the recess turned in the wood, the face-plate with chuck attached is to be taken from the lathe and laid on the bench, and the cylinder-cover driven into the recess by a few light blows from a mallet, taking care to place the cover with its outside towards the wood, so that the inside may first be turned. As to the process of turning, light cuts only should be taken, and the centre of the cover should be left projecting about $\frac{1}{16}$ of an inch, the projection being of such a diameter as to fit the bore of the cylinder exactly.

When this side is finished, the cover must be taken out of the chuck, and a recess turned in the latter for the reception of the projection in the centre of the inside of the cover; the cover is now to be replaced in the wooden chuck in the same manner as before, but in a reversed position, for the turning up of the outer surface.

To do this, turn up the flat part first, and slightly mark upon it with the graver a circle, $1\frac{1}{4}$ inch in diameter. Without removing the piece from the chuck, draw a line through its centre, and where this line cuts the circle just mentioned, centre-punch and drill two holes, $\frac{3}{32}$-inch in diameter. Now pass two $\frac{1}{2}$-inch wood screws through these holes, and screw

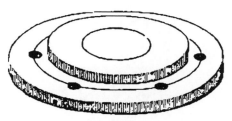

FIG. 51.

firmly down to the wooden chuck, after which the centre and hollow portion of the cover may be turned. When the surface is finished, the chuck itself must be turned away until the edge of the cover is free, and then, the cover being still held to the chuck by the two screws, the edge is to be turned down until it exactly fits the cylinder-flange. Fig. 51 shows the inner surface of the finished cover.

The top cover is to be turned up in much the same manner, the under side being turned first. Of course a deep recess to hold the stuffing-box (which is cast on to the top of this cover) must be turned in the chuck in the first instance. The top being reversed in the chuck, the stuffing-box will be turned up after the flat part has been turned and the piece screwed down to the chuck.

Now while the cover is still in the lathe, drill a hole $\frac{5}{32}$-inch in diameter through the centre of the stuffing-box, and enlarge this to $\frac{5}{16}$-inch down to within about $\frac{1}{8}$-inch of its inner end. Tap the stuffing-box with a tolerably fine screw-thread by means of a chaser, if the reader has had sufficient practice in the use of this tool, otherwise the thread must be cut with taper, medium, and plug taps. The top cover will now be finished, with the exception of the screw-holes. To drill these, hold the cover in place upon the cylinder, and mark upon the latter the position of the two screw-holes already drilled through the cylinder-cover; centre-punch and dril them through the flange with a $\frac{1}{16}$-inch drill, tap them, and screw the cover on with $\frac{3}{32}$-inch screws. Now divide the circle marked on the cover into three equal parts on each side of the two screws; centre-punch at the four points of division thus found, and drill right through both cover and flange with the $\frac{1}{10}$-inch drill; unscrew the cover and tap the four holes in flange, and enlarge those in cover to $\frac{3}{32}$-inch, and the cover will then be ready to screw on. These instructions apply equally to both top and bottom covers.

STEAM-CHEST.—In the set of castings now under consideration, the gland for the cylinder stuffing-box is cast in one piece with the gland for the slide-valve chest, and I shall therefore leave the fitting of it until after the valve-chest has been finished; and this will be our next care. The edges of the flange should first be filed up square to fit the cylinder-face. If the valve-chest has no chucking tenon (and it has none in the set of castings now before me), the centre of the stuffing-box must be found and marked, and then the chest being laid on the surface-plate or lathe-bed, a scriber must be set to this mark, and the chest being reversed, a line is to be drawn across the back of it with the scriber. The chest is now to be set up on the edge of its flange, first on one side and then on the other, each time setting the scriber to the centre of the stuffing-box, and marking the back of chest. Centre-punch the position thus found upon the back, and mount the chest between the lathe-centres. The stuffing-box is now to be turned up, and the chest being then taken from the lathe, a $\frac{3}{32}$-inch hole is to be bored through the stuffing-box, enlarged to $\frac{1}{4}$-inch diameter to within $\frac{1}{8}$-inch of the inside of the valve-chest, and screwed in the same manner as the cylinder stuffing-box.

STUFFING - BOXES.—Now take the piece from which the two glands are to be formed, and centre-mark and drill a $\frac{3}{32}$-inch hole right through it. This hole may at once be enlarged to $\frac{5}{32}$-inch through that portion of the piece which is to form the gland for cylinder-top. Mount the piece between centres with a carrier upon one end, and turn up the other end to the full diameter of the bottoms of the threads of stuffing-box it is to fit; now reverse the piece in the lathe, so that the carrier may be placed upon the end already turned, and turn the other end to fit the other stuffing-box. The flanges of the glands should be turned to a diameter about $\frac{1}{4}$-inch larger than their respective stuffing-boxes, and a portion of the metal between the two glands may at the same time be turned away, but care must be taken not to separate the glands which must next be screwed to fit the stuffing-boxes. Now screw the valve-gland into its stuffing-box as far as it will go, and mount valve-chest and glands between lathe-centres, the valve-chest acting as a carrier; mill the flanges with a milling tool, and cut off the cylinder-cover gland, turning up its face at the same time. When this gland has been cut off, bring the back centre up again, and cut off the superfluous piece of metal left

Horizontal Slide-Valve Engine. 69

on the top of the valve-chest gland, and the glands will then be finished, appearing as in Fig. 52.

FIG. 52]

The steam-chest may now be filed up square and smooth all over; the holes for the screws should be drilled $\frac{1}{8}$-inch from the edge of the flange, and $\frac{3}{16}$ to $\frac{1}{4}$ of an inch from each end. One hole $\frac{3}{32}$-inch in diameter should first be drilled through the flange of the chest, the position of this hole marked upon the cylinder-face, and a hole drilled therein with a $\frac{1}{16}$-inch drill; this hole is to be tapped, and the valve-chest screwed to the cylinder-face and carefully set square with the cylinder, so that the piston-rod and valve-rod when in place may be exactly parallel. The other three holes are now to be drilled with a $\frac{1}{16}$-inch drill through both valve-chest and cylinder-face, being afterwards enlarged to $\frac{3}{32}$-inch in the former, and tapped with screw-thread in the latter, as in the case of the cylinder-covers.

CHAPTER V.

HORIZONTAL SLIDE-VALVE ENGINE (continued).

PISTON — SLIDE-VALVE — ECCENTRIC — BED-PLATE — CRANK-SHAFT — CONNECTING-ROD — GUIDE-BARS — STANDARDS — FITTING.

THE PISTON.—This part may now be turned, as it might have been at any former stage after the boring of the cylinder was finished. If there is a chucking tenon on the piston, it may be turned up and bored and screwed for piston-rod while in the lathe. For the piston-rod, a piece of straight steel about $\frac{3}{32}$ of an inch in diameter and 5 inches long may be used, and this should be screwed into the piston while the latter is still in the lathe, so that we may be sure it is screwed in straight and square.

If, however, the piston has no tenon-piece, it may be bored, the piston-rod screwed in, and the end of the rod centre-punched, when the piston may be turned up between the lathe-centres, the driving carrier being placed on the piston-rod. The piston when finished should be a good fit in the cylinder. It should be $\frac{1}{4}$ of an inch thick, and a groove $\frac{3}{32}$ of an inch wide and $\frac{3}{32}$ deep should be turned in it to take the packing.

Another method of making the piston will now be described—a method which requires that the piston should be made in two parts: one part as shown at A, Fig. 53, the other part forming the bottom plate shown at B in the same figure. A cast gun-metal ring will be required, having an inside diameter rather less than the diameter of the groove in the piston, and an outside diameter rather larger than the diameter of the cylinder. (In larger engines two rings are used, these being bored eccentrically; and when the cylinder is of brass or gun-metal, the rings are

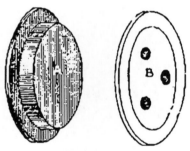

Fig. 53.

of cast iron.) The two parts of the piston must be screwed together with three small screws through the bottom plate. Of course screws with bevelled heads should be used, so that they may be driven in flush with the surface of plate; the piston-rod is then to be screwed in, and the piston turned up as already described.

The ring must be chucked in a recess turned in a

piece of wood screwed on to the face-plate, in the manner adopted for chucking the cylinder-covers. The inside of the ring is to be turned out to the same diameter as that of the groove in the piston. The sides must also be turned up square, the ring being reversed in its wooden chuck so as to get at the other side. Thus far finished, the ring should be of just such a width that when tried in its place on the piston, with the bottom plate screwed on, it will be a good fit, not jammed immovably, and not having any shake. Now put the ring on the piston and screw on the bottom plate, with a ring of paper between it and the piston ring, so that the latter may be tightly held. Place the whole between the lathe-centres, and turn down piston and ring, leaving the former a good fit to cylinder, and the latter a tight fit to same. Now take out the ring, and saw through it at an angle of 45°. Replace the ring on the piston and screw up again,

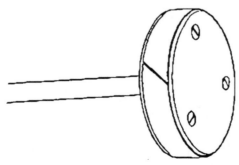

FIG. 54.

and the piston is finished, the spring of the ring keeping it pressed against the sides of the cylinder, and thus obviating the necessity for any other packing. Fig. 54 shows this form of piston complete. When two rings are used, the saw-cuts are placed upon opposite sides of the piston, thus preventing any loss of steam through either division.

A further development of this sort of piston is one in which steam is admitted to a groove in the piston under the rings, the pressure of the steam thus helping to make the piston steam-tight. This form is valuable where the piston goes on working when the steam is shut off, as in a tram engine descending an incline; for the friction of the piston against the sides of the cylinder is thus diminished directly the steam is shut off, and any unnecessary wear of these parts is consequently avoided.

Before leaving this subject, I may mention that the following proportions are given by Molesworth for the piston and piston-rod of a high-pressure engine:—depth of piston, ·25 of diameter of piston; diameter of piston-rod, ·15 of diameter of cylinder.

THE SLIDE-VALVE.—This part must be filed up square and smooth on the outside, and the width reduced until it will work easily in the slide-valve chest. The interior of the valve should be cut out with a small chisel until it measures nearly the same in length as the distance between the inner edges of the steam-ports. The metal must of course be left of equal thickness at both ends when the valve is finished, and the total length of the valve should be such that when the valve is placed over the ports it will extend a very little beyond the steam-ports at each end. This extension beyond the steam-ports is technically called "the lap of the valve," and serves to cut off the supply of steam rather earlier in the stroke of the piston than would otherwise be the case. The valve-face should be rubbed down flat on the stone already used for the cylinder-face, and a slit rather less than $\frac{1}{16}$ of an inch wide, and $\frac{3}{16}$ of an inch deep, sawn in the back of the valve lengthwise, as shown

FIG. 55.

in Fig. 55. Fig. 56, giving a sectional view of the valve in position over the steam-ports, will make the foregoing description quite clear.

THE ECCENTRIC AND ITS STRAP.—The eccentric must be chucked by the tenon, which should *always* be cast upon this piece; the edge is to be turned down smooth, and a groove, $\frac{2}{32}$ of an inch wide and $\frac{1}{16}$ of an inch deep, turned in it. While still in the lathe the true centre of the eccentric should be marked, and the chucking tenon may then be cut off.

A moment's reflection will make it apparent that when the slide-valve is in action, the port in it must cover the opening of the exhaust-port in the cylinder-face during the whole of each stroke, while, during each half-stroke, one or other of the steam-ports must also open into the valve-port, the two steam-ports each in turn becoming an exhaust-way; while, on the other hand, one or other of the steam-ports must each half-stroke be left uncovered by the valve, and this will be accomplished by making the whole travel of the valve equal to twice the width of one steam-port added to twice the lap given to the valve.

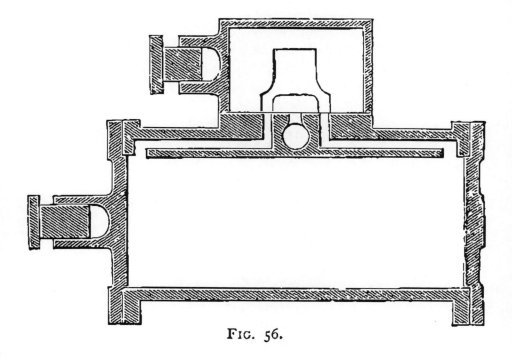

FIG. 56.

Returning now to the eccentric,—a distance equal to half the travel of the valve must be marked off from the centre of the eccentric, and as nearly as possible in the centre of the boss cast on the eccentric; this mark must be centre-punched, and a hole drilled through it of any convenient size, not exceeding the diameter of the shaft upon which the eccentric is ultimately to be mounted. The piece can now be mounted on a temporary shaft, and the boss neatly turned up.

The strap of the eccentric should now be either filed up smooth, or it may be bored first and then turned up in the lathe, being mounted for the latter purpose upon a wooden mandrel; in either case it must be finished to just the thickness that will make it a good fit in the groove of the eccentric. The band is to be bored out the same diameter as the eccentric, the diameter of the latter being of course callipered

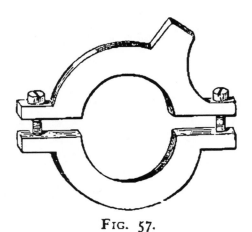

FIG. 57.

at the bottom of the groove. The outside edge of the band must be smoothed off with the file, and $\frac{1}{16}$-inch holes bored in the two projections or lugs, and tapped for screws. The band is now to be sawn in two, as shown in Fig. 57, and the holes in the upper half slightly enlarged to allow the screws to pass through.

THE BED-PLATE.—If this has not been purchased, it may now be cut out; a convenient size will be found to be $10\frac{3}{4}$ inches long and $3\frac{5}{8}$ inches broad. Half an inch from one end a rectangular hole $1\frac{1}{2}$ by $1\frac{1}{4}$ inch must be cut out, as shown in Fig. 26, in order to allow the crank to revolve. As to the thickness of the bed-plate, the metal used should be $\frac{1}{8}$ or $\frac{3}{32}$ of an inch thick.

THE CRANK-SHAFT.—This part is made from a forging, which may be obtained at any of the shops that supply castings, for the sum of 4*s*., or from any country blacksmith for about 6*d*.; in fact, the latter will turn out the forging "while you wait." All that is necessary is to get him to weld a piece of $\frac{3}{8}$-inch iron $1\frac{1}{4}$ inch square on to a piece of $\frac{3}{8}$-inch bar,

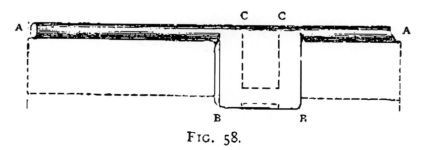

FIG. 58.

and your 4*s*. forging is ready, as shown in Fig. 58. The dotted lines C C in this figure indicate that part of the forging which must be filed away—a piece of work which, it may be remarked, is not to be accomplished in a few minutes. As some amateurs with more money at command and less time than others object to this labour, Messrs. Lucas & Davies have thought it best in these days when most things are made easy, to bring out a bent crank which only requires turning and finishing off. However, the task of filing away the solid crank forging may be rendered less tedious by drilling two or three holes through the part to be removed, which reduces considerably the amount of labour. To turn the crank-shaft it must be placed between the lathe-centres, with a piece of hard wood in the gap, and turned down to ¼-inch diameter, the proper proportion being ·33 of the diameter of the cylinder. The bearing-pin of the crank will be more difficult to turn up. When the shaft is a short one, and the lathe-centres are sufficiently long, the pin may be centre-punched and mounted between the centres, and the pin turned down.

The following is also a simple method of turning the crank-pin:—To the ends A A of the shaft solder two flat pieces of iron rather longer than the throw of the crank ; now insert between the ends of these pieces of iron and the crank B B, two pieces of hard wood to take the thrust of the lathe-centres, in the same manner as was done by the piece of wood in the gap of the crank while the shaft itself was being turned. On the pieces of iron mark off the distance from the centre of the shaft equal to the throw of the crank, centre-punch and mount between the lathe-

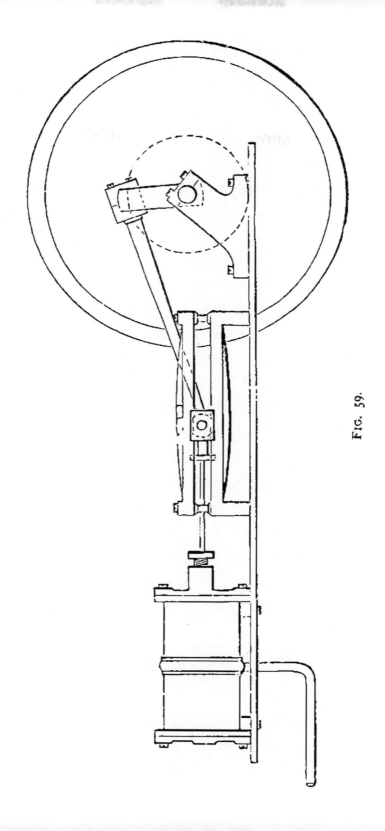

Fig. 59.

centres, when, if the irons have been soldered on in a line with the crank, the crank-pin will be between the lathe-centres, and may easily be turned up.

The pin when finished should be about $\frac{5}{16}$ of an inch in diameter, or ·23 of the diameter of the cylinder. The flat part of the crank should be finished with the file.

Before proceeding further, it will be best to make full-size working drawings from which to fit the several parts of the engine together.

Figs. 59 and 60 are working drawings on a scale of 6 inches to the foot. It will be seen that the parts are so arranged that the length of the connecting rod is just double the stroke of the piston.

THE CONNECTING-ROD will therefore be 4 inches in length, in the case of a 2-inch stroke engine; if it is not provided in the set of castings, it may be made from a round bar of iron, the ends being forged out flat, or a flat piece of iron half an inch wide and a quarter thick may be taken, the middle part filed roughly to shape, and then turned down, as shown in Fig. 61. Drill the top end of the

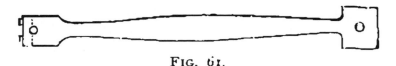

FIG. 61.

connecting-rod for the crank bearing, and also drill two screw-holes in the top; now saw across, as shown by the dotted line in Fig. 61, and enlarge the screw-

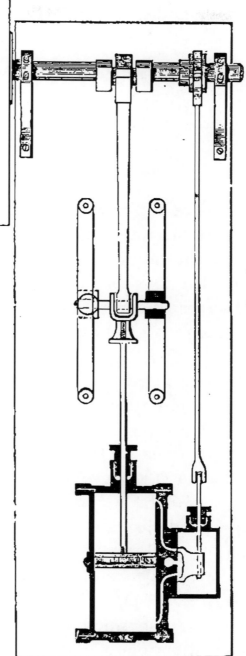

Fig. 60.

holes in the top piece and tap those in the rod itself for the screws which are to fasten the two parts of the bearing together. Another way of making the bearing in the connecting-rod is shown in Fig. 62; the

FIG. 62.

top piece A may be filed up from a piece of steel, and then fitted to the connecting-rod, and a $\frac{3}{32}$-inch hole drilled through as shown by the dotted lines in B. A pin is to be filed to size and run through the hole, and the two parts being thus firmly held together, the hole seen in B is to be drilled to fit the crank.

The different parts of the engine should now be fitted together, each piece being finished and fitted as required. If the bed-plate has been bought with the castings, the positions of the various parts will be found marked upon it; but if the bed-plate is a home-made one, it will be necessary to set out the positions of the cylinder, guide-bars, etc., from the working drawings.

The lugs on the cylinder must be bored, and two pieces of $\frac{1}{8}$-inch steel about $\frac{5}{8}$-inch long must be screwed into them. The position of these bolts relatively to the top and bottom of the cylinder must be taken with the compasses and marked on the bed-plate, and a couple of holes drilled through the latter to allow the bolts to pass through. A couple of nuts screwed on to the bolts will hold the cylinder firmly down in its place. Should it prove to be not perfectly level on the bed-plate, it must be removed, and the lug at the high end filed away until it is found that the piston-rod is perfectly parallel with the bed-plate when drawn out to its full extent. The pillars to go between the guide-bars should now be turned up; this is done while they are all in one piece. They are afterwards sawn apart, when their height should be just a shade over the thickness of the guide-blocks, which are in the present instance $\frac{1}{4}$-inch thick. They must be bored with a $\frac{1}{8}$-inch hole, and this may be done to each pillar separately, or to all four at once while they are in one casting.

THE GUIDE-BARS.—These having been filed up smooth, and the upper surface of the lower bars and the under surface of the top bars worked perfectly level, the top guide-bars must be centre-punched at each end and drilled with a $\frac{1}{8}$-inch hole. They are then to be placed upon the lower bars, and the position of the holes in the top bars marked, centre-punched and drilled in the lower bars. These latter must now be placed in position upon the bed-plate, and the position of the holes again marked and drilled in this latter. Both bed-plate and bars should be marked so that they may always be put together in the way in which they were marked for drilling. Before putting the guide-bars together the lower ones may be roughly filed down to near the correct height; and as the distance between the guide-bars will be $\frac{1}{4}$-inch, and the piston-rod will work in the centre of them, it is obvious that the top surface of the lower guide-bars must be $\frac{1}{8}$-inch nearer the bed-plate than the centre of the piston-rod. Four bolts about $1\frac{3}{4}$-inch long, with small steel nuts, will be required to bolt the guide-bars to the bed-plate. These bolts are passed through the guide-bars with the pillars between them, and the whole thus secured in position.

The Crank-shaft Bearings or Standards may next be finished off. A $\frac{5}{16}$-inch hole is to be bored in each standard, and two screw-holes drilled in the top of each to the depth of half an inch, and tapped for $\frac{1}{8}$-inch screws. Then the tops are to be sawn off at the lines marked on the castings, and the holes in the top pieces enlarged to allow the screws to pass through them. The feet of the standards must also be drilled to take $\frac{1}{8}$-inch bolts or screws, the former being preferable. The crank-shaft may now again be put in the lathe, and the parts which are to rest in the bearings turned down to fit the holes in the latter. The fly-wheel end will only be turned down for a length equal to the thickness of the bearing; at the other end, however, it will be a good plan to turn the shaft down to the same diameter to within half an inch of the crank, as the hole in the eccentric will not then need to be larger than those in the standards, and the small eccentric supplied with model castings scarcely admits of a larger one.

The eccentric should be bored out to be a good fit in its place on the shaft, and a small hole is to be bored through the boss upon it sloping inwards into the larger one; this is for the set-screw, and should be tapped for $\frac{1}{8}$-inch screw. The set-screw itself is to be made from a piece of steel wire $\frac{3}{8}$-inch long, and screwed and finished with a nick in the top for the screw-driver, no head being necessary.

The crank-shaft is now to be placed in the bearings, and the whole placed upon the bed-plate. Now see that the shaft is directly at right angles with the piston-rod, and then mark the positions of the holes in the feet of the standards upon the bed-plate. Drill the latter and bolt the standards down.

THE PISTON-ROD HEAD is to be bored and tapped with a screw, so that it may be screwed on to the piston-rod when the proper thread has been cut upon this latter. A $\frac{3}{16}$-inch hole must be bored through this piece to take the cross-head. The piston-rod should be cut off to the right length, which length may easily be taken from the working drawing, and screwed with the same thread as that of the piston-rod head, which should now be fitted on to it.

Attach the connecting-rod to the crank, and after pushing the piston down to about $\frac{1}{8}$ of an inch from the bottom of the cylinder, and bringing the crank round until it is at its nearest to the cylinder, mark on the free end of the connecting-rod the position of the cross-head holes in the piston-rod head. Take the connecting-rod off the crank and drill a $\frac{3}{16}$-inch hole for the cross-head in the place marked.

Take a piece of $\frac{3}{16}$-inch steel $1\frac{3}{4}$ inches long, and turn half an inch at each end down to $\frac{1}{8}$-inch. Drill a $\frac{1}{8}$-inch hole through the centre of each guide-block. Now take the top guide-bars off the bed-plate and pass the steel cross-head through the piston-rod head and connecting-rod, placing the guide-blocks on the end of the cross-head.

The height of the guide-bars with respect to the piston-rod and of the pillars between the upper and lower bars must be carefully adjusted, a very little being taken off the feet of the lower bars till these latter are low enough, or if they are accidentally made too low, the lugs by which the cylinder is bolted down must be slightly reduced; for it is obvious that unless the guide-bars are perfectly horizontal and of exactly the correct height, a strain will be put upon the piston-rod and much unnecessary friction caused, even if the piston can be got to move at all under such circumstances. At the same time the thickness of the guide-blocks must be adjusted so that the blocks will run smoothly and easily, and without any shake, between the guide-bars.

THE SLIDE-VALVE ROD is to be made of steel

wire, say $\frac{3}{32}$-inch thick, rather under than over. The eccentric band must be drilled and screwed to receive one end of the rod, and a joint must be made in the rod about one inch from the steam-chest stuffing-box. The short piece is to be filed flat on two sides $\frac{1}{16}$ of an inch from its end for a distance corresponding with the length of the valve, so that it will drop into the slit in the latter, and will move it backwards and forwards over the ports without allowing any end-play. The joint in the valve-rod may be made from two pieces of brass-wire $\frac{3}{16}$-inch thick; these are each drilled and screwed at one end, to receive the ends of the two parts of the rod; in the other end of one piece a saw-cut is to be made, and a flat filed on the other piece to fit the saw-cut. These two pieces are now to be put together, the flat of the one into the saw-cut of the other, a small hole drilled through the joint, and a piece of wire put through as a pivot, and rivetted with one or two light taps of the hammer.

The eccentric must now be put in place on the shaft, and the length of the rod so adjusted that when the strap is at its farthest from the steam-chest, the lower port shall be just as much open as the upper port is when the strap is at its nearest to the steam-chest; of course for this purpose the steam-chest must be taken off, and the engine turned on its side with the cylinder-face uppermost, so that the valve may lie upon this latter.

The fly-wheel may be turned and bored upon the face-plate of the lathe, or it may be keyed on to the crank-shaft and turned up in position. The shaft must be filed away a little on one side, and a corresponding flat filed in the fly-wheel; a small key very slightly wedge-shaped is then to be driven into the space thus made between the wheel and shaft, bearing in mind however the caution given in Chap. II.

An exhaust-pipe may be screwed into the exhaust-port of the cylinder, brought through a hole in the bed-plate and bent outwards, and a steam-pipe should be screwed into the steam-chest.

If it has not already been done, the burrs left at the edges of the screw-holes in the cylinder-covers, steam-chest, etc., must be rubbed off, and a little white and red lead put between such parts as are required to be steam-tight before they are finally screwed together. Before the steam-chest is screwed on, however, the eccentric must be fixed by means of its set-screw in such a position upon the shaft that the valve will open the lower port just as the piston is at the bottom of the cylinder, and the upper port just as the piston is at the top of the cylinder.

Oil-holes may be drilled in the centre of the top guide-bars, the plummer-blocks, and anywhere else where it may be thought necessary.

The engine may now be mounted upon a solid block of wood, the same size as the bed-plate and $2\frac{1}{2}$ inches thick, holes being made in the top of the block to

FIG. 63.

make room for the various nuts under the bed-plate, and a slot cut for the exhaust-pipe; and the bed-plate being screwed down to this block, we may consider our model finished.

Fig. 63 shows the engine in its complete state, but mounted upon standards instead of upon a solid block of wood. From this diagram the relative disposition of the different parts of which it is composed will be easily recognised.

The amateur who has successfully made this model may safely take in hand a small power engine of the class recently introduced by Messrs. Lucas & Davies; these engines are very suitable for driving a lathe, working a pump or a small circular saw, or other machinery likely to be found in the amateur's workshop.

The cylinder is of iron, and if the cylinder and guide are bored before the castings are sent out, the engine will be an easy one to put together. The set, including governor and force-pump castings of the No. 1 engine, cylinder 1½-inch diameter and 3-inch stroke, costs 30s.; while the set for the No. 2 size, cylinder 2-inch diameter and 4-inch stroke, costs 50s. The nuts and studs for No. 1 are sold at 12s. 6d., and for No. 2 at 15s. per set. Tubular boilers suitable for these engines, fitted with pressure-gauge and all necessary taps, etc., are supplied by the same firm at £10 10s. and £13 respectively. The finished engines complete with boiler, etc., cost £20 and £25.

CHAPTER VI.

VERTICAL SLIDE-VALVE ENGINE.

THIS engine is the same in all essential respects as the one last described, the only difference being in the arrangement of the working parts vertically instead of horizontally; this type of engine consequently occupying less floor space than that last described.

The castings required for this engine consist of the following pieces :—

Two A-standards,
Two bearings,
Cylinder,
Two cylinder-covers,
Piston,
Steam-chest,
Slide-valve,
Gland for steam-chest,
Gland for top of cylinder,
Eccentric and eccentric-strap,
Support for guide-bars,
Cross-head,
Crank,
Fly-wheel.

Some makers will also supply the connecting-rod as a casting.

Of these castings, the cylinder, steam-chest, piston, glands, connecting-rod, slide-valve, eccentric and eccentric-strap, and fly-wheel, will be in all respects similar to the castings of those parts used for the horizontal engine; and as they are to be treated in the same manner up to the point of fitting the engine together, we shall leave them on one side for the present, and proceed to speak of the pieces which vary in this engine from the analogous parts in the horizontal type.

These pieces are—the guide-bar stay, shown in Fig. 64; the cross-head, Fig. 65; the crank, which will be of the same design as that of the oscillating engine first described; and the cylinder-cover, which differs from the horizontal cylinder-cover in having two small projections on its rim into which the guide-bars are fixed.

The bed-plate is shown in Fig. 66.

This engine makes a very pretty model, and with very little extra trouble it may be brought out highly finished, all the parts being filed smooth and burnished, with the exception of the body of the cylinder, which may be coloured with bright green enamel paint.

The surface of the bed-plate should be filed up flat and smooth; the position of the centre of the cylinder must be marked upon it; and when the

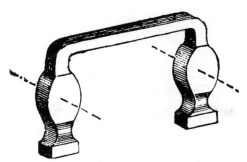

Fig. 64.

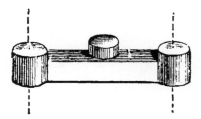

Fig. 65.

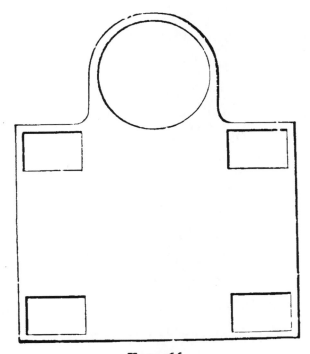

Fig. 66.

circle for the screw-holes has been marked upon the cylinder-bottom, a circle of the same radius should be struck upon the bed-plate from the centre marked, and two holes of a size to allow the screws to pass through them should be drilled diametrically opposite each other upon the circumference so struck. The cylinder-bottom must then be placed upon the bed-plate, and while it is held in place, mark upon it the position of these two holes just mentioned; the position of the other four holes for screwing on the cover being marked off as usual. When the bottom cover is screwed on to the cylinder, the two holes coinciding with those in the bed-plate are to be left without screws, to be used in finally attaching the cylinder to the bed-plate. It will give a neater appearance to the engine, when finished, if screws with conical heads are used for screwing on the cylinder-bottom,* the screw-holes being countersunk for the reception of the heads of the screws, or all the six screws may be put through both bed-plate and cylinder-cover.

The two projections on the cylinder-top will prevent its edge being turned up in the lathe; therefore when each surface has been turned up, and before the top is taken from the lathe, a circle should be marked with the graver of the same diameter as the outside of the cylinder-flange, and the edge must afterwards be carefully filed away to this mark, save where the aforesaid lugs are left projecting. In marking the screw-holes upon the top flange of the cylinder, it must be remembered that when the engine comes to be finally fitted up, a line drawn through the lugs on the cylinder-top must be at right angles with the shaft of the engine.

The standards should be filed up smooth and the tops left level; the ornamental grooves may either be scraped out bright or left as they are, or they may be painted over with enamel paint after the engine is finished.

The bearings may next be filed smooth, centre-punched, and drilled for screws; they must then be put in place upon the standards, and the position of the screw-holes marked on these latter, and drilled and tapped for $\frac{1}{8}$-inch screws. The bearings should then be screwed firmly to the standards and filed down to the same level at the sides and ends while *in situ*. If properly done, the joint should be invisible except upon looking closely for it. A groove is left in the castings for the bearings, and in the tops of the standards, so that when the two parts are screwed together the hole for the crank-shaft will be formed between them. To bore this hole out to the right size, a small D-bit is the best tool. The finished diameter of the hole should be $\frac{1}{4}$-inch.

Before the bearings are taken off the standards, they should be marked at the back, so that they may always be placed in their proper position without trouble. The feet of the standards may now be drilled for the bolts which are to hold them down upon the bed-plate.

The crank may be simply filed up smooth, or the bosses for the end of the crank-shaft and for the crank-pin may be finished off in the lathe.

The support for the guide-rods will be fixed in the position shown in Fig. 68. A hole must be drilled in each of the two legs of the front standard, and the guide-bar support being drilled and tapped in correspondence with these holes, is held in place by screws passed through from the back of the standards.

The centres of the two bosses shown in Fig. 64 should now be found and punched. The exact distances between these centres must be taken with the compasses and marked off upon the cross-head (Fig. 65), and upon the cylinder-top. These pieces are then to be drilled as marked, the holes in the cover must be tapped to take a piece of screwed steel rod about $\frac{1}{8}$-inch in diameter, the cross-head should be an easy fit upon the same rod, so that it will slide up and down without shake, and the holes in the support must be the least trifle smaller than those in the cross-head, so that they will be a tight fit upon the same rods. The holes in the support may be drilled right through or not, according to fancy; in the former case, the guide-rods will of course be made rather longer, their tops being neatly rounded off. The cross-head must be drilled and tapped for the end of the piston-rod, and fitted to the connect-

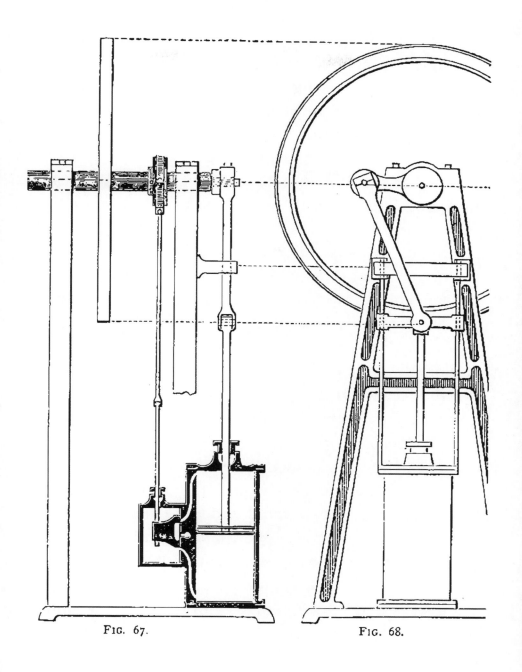

Fig. 67.   Fig. 68.

ing-rod, the joint being made by a steel pin passed through the connecting-rod and the cross-head, the other end of the connecting-rod being treated in the manner described in the case of the horizontal engine.

The shaft will be of the same diameter as that used for the horizontal engine; it will be made of a straight piece of round steel bar turned down to fit the bearings, and $\frac{3}{8}$ of an inch at the cylinder-end must be turned down to $\frac{1}{4}$ of an inch diameter and screwed, the larger boss on the crank being drilled and tapped to correspond. The smaller boss on the crank is to be drilled and tapped to take a $\frac{3}{16}$-inch screw, a steel pin one inch long being screwed into the same.

The engine may now be put together. Let the feet of the standards be filed down, if necessary, until the shaft-bearings are perfectly level with one another and the shaft runs easily. The position of the eccentric will of course be inside the front standard, and the cylinder, connecting-rod, fly-wheel, etc., being all in place, the position of the holes in the feet of the standards must be marked upon the bed-plate, and the cylinder being unscrewed, the bed-plate is to be drilled where marked; the standards can then be bolted down, and the engine permanently fixed up.

Figs. 67 and 68 are working drawings on a scale of six inches to the foot, Fig. 67 showing the engine partly in elevation and partly in section. In this figure, however, the distance between the steam-ports and the exhaust is greater than it should be.

CHAPTER VII.

DOUBLE-CYLINDER LAUNCH ENGINE.

THE engine described in this chapter is a double-cylinder launch engine, the castings for which cost 14*s*.

The set consists of thirty-three pieces, as follows:—

1 Bed-plate (Fig. 71),
1 Top-plate (Fig. 72),
4 Standards,
2 Cylinders,
2 Cylinder-tops,
2 Cylinder-bottoms,
2 Pistons,
2 Steam-chests,
2 Slide-valves,
2 Connecting-rods,
2 Eccentrics,
2 Eccentric bands and rods (Fig. 69),
2 Cross-heads,
3 Shaft-bearings,
1 Balance-wheel,
1 Screw,

and two pieces, each of which forms a cylinder and steam-chest stuffing-box gland.

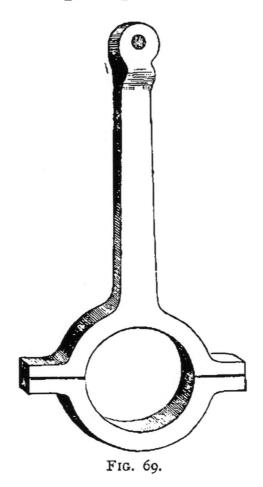

Fig. 69.

The cylinders are 1 inch in diameter, and $1\frac{1}{2}$ inch stroke.

The steam-chests, cylinder-bottoms, and eccentrics, all have chucking tenons cast upon them for convenience in turning.

The cylinders, with their steam-chests and slide-valves, will of course be finished off in the same

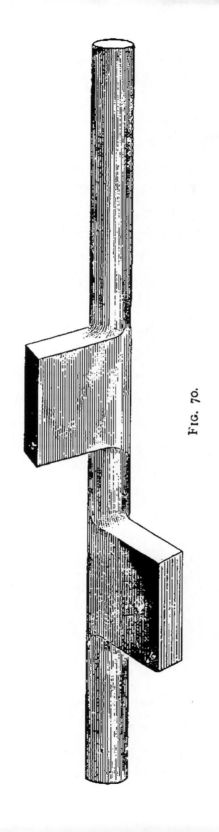

FIG. 70.

manner as in the other engines described, except that the cylinders will have to be bolted down upon the top-plate. This may be done by making screw-holes in the plate to coincide with those in the cylinder-tops and cylinder-flanges, and screwing plate and covers to the cylinders with the same screws. Three marks will be found on the top-plate, one showing the centre of the plate, and the other two indicating the position of the centre of each cylinder. Taking these two latter marks as centres, two half-inch holes must be bored in the top-plate for the stuffing-boxes of the cylinder-covers to pass through.

The standards are best bolted to the bed-plate with small bolts and nuts, using two for each standard. The positions of the standards are shown at A (Fig. 71). The top-plate is placed symmetrically upon the tops of the standards, and secured in the same manner.

The crank forging is made as described in a previous article, except that in this instance there must be two crank-pieces welded on to the bar, and these pieces should be at right angles to each other, as seen in Fig. 70.

The cross-heads must be filed up, and a slight groove filed in each end to fit the guiding part of the standards.

The fitting of the connecting-rods to the cross-heads and cranks will present no difficulty, the cross-head and connecting-rod being held together by a $\frac{1}{8}$-inch steel pin ; and the crank end of the connecting-

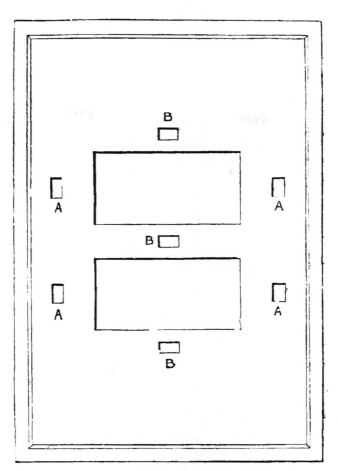

FIG. 71.

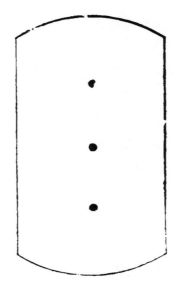

rod being sawn across where marked, the two parts are screwed together and drilled to take the crank.

The position for the three supports for the shaft is shown at B (Fig. 71). They must be sawn across, and drilled with a ¼-inch drill for the shaft, the latter being turned down to the same diameter at the parts which are to rest in the bearings.

The eccentrics and eccentric-bands are to be fitted as in the other engines already described; half the band will, however, be in one piece with the eccentric-rod. In the top of the eccentric-rod, a slot must be filed to receive the head of the slide-valve rod, which latter is formed from a piece of brass rod filed flat at one end; the two parts are then put together as a hinge, and held by means of a small pin.

The bed-plate must, if necessary, be filed away to make room for the revolution of the eccentrics.

The balance-wheel, answering to the fly-wheel of other engines, must be screwed or keyed on to the end of the shaft.

The screw must be filed up and mounted on a separate shaft, the length of which will depend upon the boat in which the engine is to be placed. The shaft is made from a piece of steel wire; the screw is drilled, and a set-screw fitted into it upon one side to fix it in its place upon the shaft; upon the other end of the shaft is screwed a piece of brass, through which a pin passes, which engages two pins projecting from the disc of the balance-wheel of the engine. The shaft passes through a long stuffing-box (as shown in Fig. 76); this latter, being firmly fixed in the hull of the vessel for which

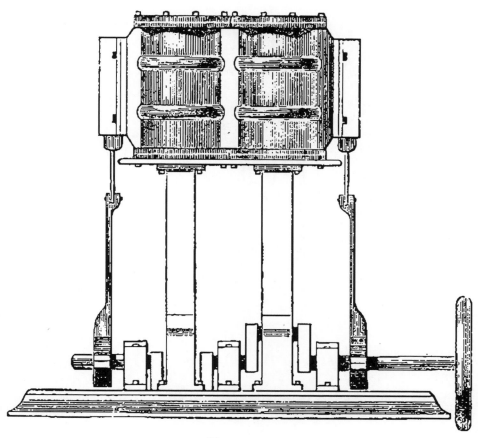

Fig. 73.

the engine is intended, acts also as a bearing for the shaft.

In order to keep the driving-pin up to the balance-wheel, a small brass collar is fixed upon the shaft; for the sake of clearness, this collar is shown in the drawing pulled forward, but in action it is fixed to

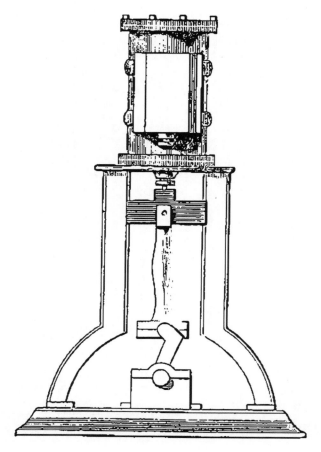

Fig. 74.

the shaft by means of a set-screw, and revolves with it against the forward end of the stuffing-box.

Fig. 73 shows the complete engine as seen from one side; Fig. 74 is an end view showing one cylinder, with its guide and connecting-rod; and Fig. 75 is the same view, with the eccentric and eccentric-rod, and slide-valve rod added.

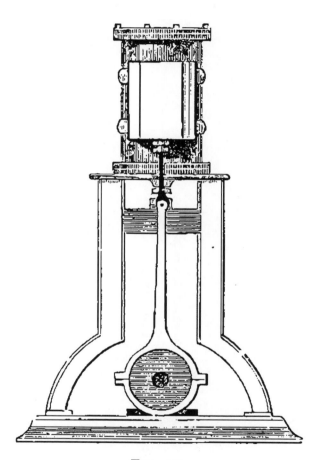

FIG. 75.

Reversing gear may easily be added to this engine; two small brackets attached to the standards carry the rocking-bar, and the reversing lever may be pivoted to a bracket attached to one of the standards

108 *Double-Cylinder Launch Engine.*

on the opposite side to those carrying the rocking-bar. A full description of reversing gear is given in another chapter.

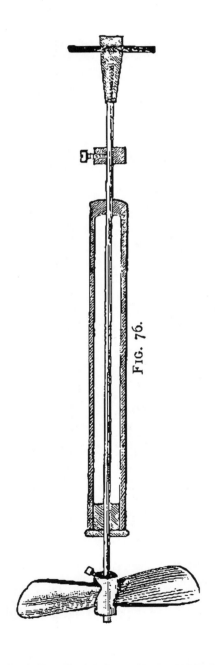

Fig. 76.

CHAPTER VIII.

DOUBLE-ACTION OSCILLATING CYLINDER MARINE ENGINE.

THE design of the marine engine in actual use in steam-ships is rather too complicated to form an easily made model, neither would a model so made be likely to prove so effective as the more simple pattern generally put into model steamers.

The full-sized marine engine is usually of the compound high-pressure type; that is, it contains within itself the essential parts of a high-pressure and of a low-pressure condensing engine, the steam being first used in a small high-pressure cylinder, and after doing a portion of its work therein it is conducted into a larger cylinder with condenser and air-pump attached, where it completes its work expansively.

In these engines the steam is admitted into the cylinders by means of slide-valves; and where the cylinders oscillate, as is often the case, the slide-valve chest is bolted to the cylinder and rocks with it, and various arrangements are used for the purpose of connecting the eccentric with the oscillating valve. In the case of oscillating cylinders the steam is admitted to the valve-chest through one of the trunnions, which are hollow and connected with the steam and exhaust pipes respectively by means of steam-tight joints.

The model I am now about to describe is similar to one large class of marine engines in that it has oscillating cylinders; and in outward appearance it will certainly pass to the unprofessional eye as a model of the real thing. Moreover, it is easy and interesting to construct, as it differs considerably from the forms previously described, and forms a handsome little engine when finished.

A set of castings for a marine engine of this description costs 9*s.* 6*d.*, and should include the following items:—

Two plates, one forming the bed-plate and the other the top-plate of the engine;

Two cylinders, differing from the cylinders of the engines described in previous chapters in having a round plate on one side and a trunnion on the other;

Four pillars to support the top-plate,
Four cylinder-covers,
Two pistons,
Two stuffing-boxes,
Two piston-rod heads cast in one piece,
One steam-block,
Two standards,
Three bearings,
Four paddle-wheel sides,—

making twenty-seven separate castings in all.

Oscillating Cylinder Marine Engine.   111

We will suppose that the cylinders are bored, and the piston, piston-rod, covers, and stuffing-boxes are all fitted, these parts being precisely the same as in the other engines described. We shall therefore commence with the steam-block (Fig. 79). This must be filed up bright and smooth, care being taken to get the two faces square with the bottom of the block, and parallel with each other. Now mark off the centre on each side of the block, centre-punch, and drill a $\frac{3}{32}$-inch hole from each side, the holes meeting in the middle. The next thing will be to make the steam-ways, and for this we shall require a special tool. A pin-drill, with the part between the pin and the outside cutting edge cut away, would do; but in Fig. 77 is shown a tool which may be made in ten

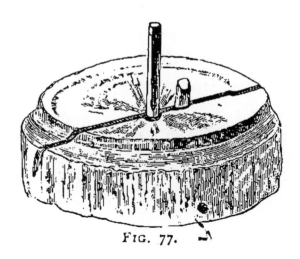

FIG. 77.

minutes, and with it both steam-ways may be cut out in another five. The writer having recently made this use of the tool from which the drawing is made, can guarantee both its speed and its efficiency. The tool is prepared in this way:—A block of box or other hard wood is to be screwed to the face-plate of the lathe, and the face of this block turned down smooth; then, while it is still in the lathe, a hole under $\frac{3}{32}$ of an inch in diameter is to be drilled in the centre. The face-plate must now be taken off the lathe, and a saw-cut $\frac{1}{4}$ of an inch deep is to be made, passing exactly through the centre of the face of the block.

A piece of straight $\frac{3}{32}$-inch steel wire must be hammered into the hole, and a cutter of the shape shown in Fig. 78 is to be filed up from a piece of sheet steel

FIG. 78.

(for example, a bit of an old saw), and after it has been hardened and tempered, this also must be hammered into the saw-cut in the block of wood, the inside edge of the cutter being $\frac{1}{8}$ of an inch distant from the centre-pin. In use the centre-pin of the tool works in the centre-hole in the steam-block, and the latter being fed up to the cutter by the screw of the lathe, a circular groove about $\frac{3}{32}$ of an inch deep will soon be scraped out.

This groove having been cut in both sides of the steam-block, a line is to be marked across each side of the block at right angles to its base, as seen in Fig. 79, and holes of the same diameter as the annular groove must be drilled through the block at the points where the line cuts the groove. Now heat a piece of brass wire of the right size for the purpose, coat it with solder, and put it through the holes just drilled; heat the block until the solder runs, and when cool, file the ends of the wire down smooth with the faces of the block.

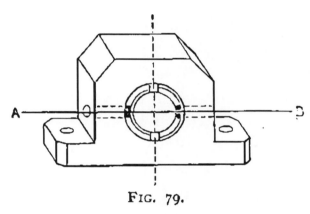

FIG. 79.

Drill a ⅛-inch hole into each end of the steam-block, ¼ of an inch deep, and two 1/16-inch holes leading from each groove into the ⅛-inch hole upon its own side of the block. Fig. 80 is a section of Fig. 79

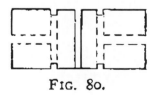

FIG. 80.

on the line A B, the dotted lines showing the steam-ways and exhaust.

After the faces of the block have been filed smooth again, a piece of 3/32-inch steel is to be fitted into the centre hole, and filed off to such a length that it will project about $\frac{1}{8}$ of an inch upon each side of the block, and the lugs of the latter must be drilled to allow the passage of the $\frac{1}{8}$-inch screws by which it will be attached to the bed-plate.

The steam-block being now complete, we will turn our attention to the cylinders. The steam-face and end of trunnion must be filed up smooth and square with the ends of cylinder; mark and centre-punch the centre of the steam-face, and draw a line from end to end of the cylinder passing through this centre mark; upon the line so drawn mark off a distance above and below the centre equal to the distance from the centre of the steam-block to the centre of the groove in the same. Centre-punch at these points, and drill three $\frac{3}{32}$-inch holes $\frac{1}{4}$ of an inch deep, one in the centre of the steam-face, and the other two at the points marked. Into these two latter holes drill steam-ways from the two ends of the cylinder, as in the case of an ordinary slide-valve cylinder. Fig. 81 shows the cylinder-face finished.

Now stand the cylinder on end upon a level surface, and mark off on the trunnion the height of the centre of the steam-face, and drill a $\frac{1}{16}$ to $\frac{3}{32}$-inch hole a little way into the trunnion. Take the two standards, and having carefully filed them up square and smooth, stand them upon a level surface and mark across them with the scriber the exact height of the centre of the steam-block. Centre-punch and

Oscillating Cylinder Marine Engine. 115

drill, and tap for $\frac{1}{8}$-inch screws, also drill the feet of the standards to clear $\frac{3}{32}$-inch screws.

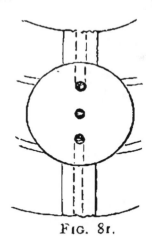

Fig. 81.

The $\frac{1}{8}$-inch screws for the standards must be filed to a smooth round point, and either lock-nuts should be fitted to the screws, or set-screws should be fitted into the top or at one side of the standards, otherwise the bearing screws will work loose with the motion of the cylinders.

The four pillars, seen in Figs. 83 and 84, may next be turned up, the top and bottom of each being screwed and nuts fitted to the bottom-pins. The top and bottom plates must be filed up quite smooth and drilled at the corners. It is best to clamp them together in the same position with respect to each other as they will occupy when finished, and drill through both at once. Use in the first place a $\frac{1}{8}$-inch drill, and screw the holes in the top plate to suit the top screws of the pillars; the holes in the bottom plate are to be enlarged to allow the lower pins of the pillars to pass through them.

Draw centre-lines across the upper surface of the foundation-plate, and placing the steam-block so that its centre coincides with the centre of the plate, mark upon the latter the position of the holes in the lugs of the block, and drill and screw the plate for $\frac{1}{8}$-inch screws.

Mark and drill the holes for the $\frac{3}{32}$-inch screws which are to fasten the standards to the bed-plate, taking care that they are in perfect line, and square with the steam-block, and also set near enough to the ends of the bed-plate to allow the cylinders to work between them and the steam-block.

The crank-shaft may now have our attention; this may be made in the manner already described, or as shown in Fig. 82, the latter form having the advantages that no special forging is required for it, and of giving a more equable motion to the paddle-wheels. The shaft itself consists of three straight pieces of steel, upon which are keyed four discs of iron or brass about $\frac{3}{16}$ of an inch thick; these discs are connected in pairs by pins upon which work the heads of the piston-rods. The drawing Fig. 82, which however is not to scale, will make this quite clear. The crank-shaft should be left of ample length, say nine or ten inches, until the width of the boat in which the engine is to work is known.

The three crank-bearings must be screwed upon

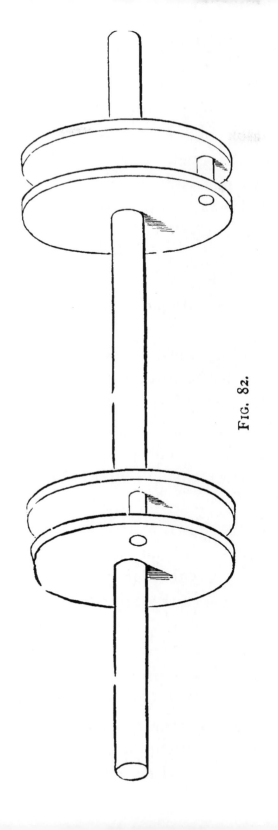

Fig. 82.

the top-plate so that the centre line of the shaft will be directly over the line running through the centres of the steam-block and the cylinder-standards, the bearings being first sawn apart and drilled for the shaft as explained in Chap. V.

The piston-rod heads must be sawn apart and drilled and screwed for the piston-rods; each head must then be sawn across, screwed together, and bored for the crank-bearing, in the same manner as the head of the connecting-rod in the other engines described.

Now the final adjustment of the cylinders should be made, the cylinder-faces being ground to the faces of the steam-block, and it is best during this process to stop up the steam-ports in the cylinder-face, otherwise the abrading material used may find its way to the interior of the cylinder, which it will in no way improve.

The engine being now finished, with the exception of the paddle-wheels, it may be screwed together, and the tops of the supporting pillars filed off level with the upper surface of the top plate.

The floats for the paddle-wheels should be about $1\frac{1}{2}$ inch in length by $\frac{1}{2}$-inch wide. The length here given is not in the proportion used for large steamers, for Molesworth gives the following as the dimensions of the paddles of one of Maudslay & Sons' marine engines: "Diameter of paddle-wheels to centre of floats, 19 ft. Length of floats, 8 ft. 6 in.

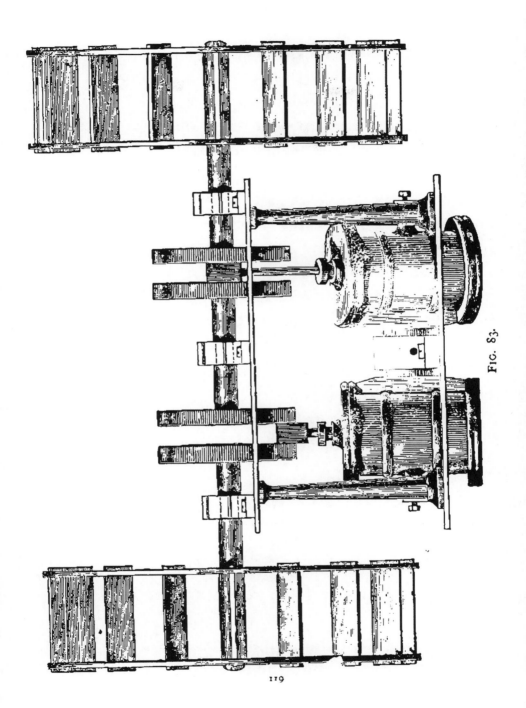

Fig. 83.

Breadth of floats, 3 ft. 4 in. ;" and the same proportion would give a length of over 2 inches to the floats of our small paddle-wheels. The floats should be made of thin sheet brass or other metal cut to size.

The sides of the wheel in hand being clamped together, the centres must be drilled to take a piece of stout tubing of a size that will just fit over the ends of the shaft; this piece of tubing should be fitted with a set-screw in the centre.

The sides must now be clamped with some small pieces of wood between them to keep them just $1\frac{1}{4}$ inch apart all the way round; the centre-tube is first soldered in, care being taken that the set-screw will come between two of the floats so that it may be reached with a screw-driver.

The floats must rest upon the bars provided for them, and will project slightly upon each side of the wheel.

A finish may now be given to the engine by painting the bodies of the cylinders green and the paddle-wheels the orthodox bright red; and if the rest of the engine has been worked up bright, one of the prettiest models we have yet constructed will be the result.

By fitting up this engine as shown in Fig. 84, instead of according to the foregoing description and illustration (Fig. 83), it may be adapted to working a screw instead of paddle-wheels.

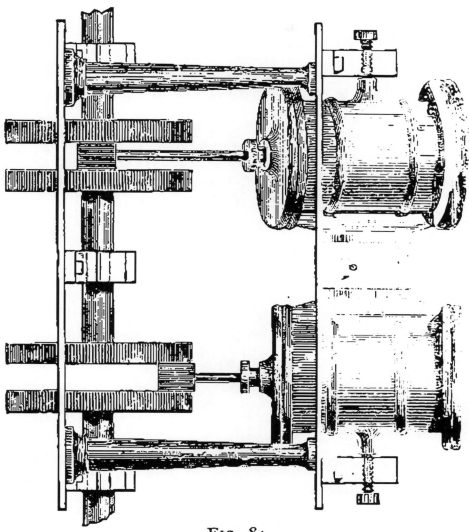

FIG. 84.

CHAPTER IX.

SLIDE-VALVE LOCOMOTIVE ENGINE.

THE locomotive, the last model to be described, will tax to the utmost the amateur's skill in fitting; but when finished, the result will be a complete and handsome model of which the maker may well feel proud.

The castings for the engine described in this chapter cost 28*s.* the set, and comprise :—

Bed-plate,
Front and back of smoke-box,
Front and back of fire-box,
Boiler,
Tubes for boiler,
Furnace-door (Fig. 95),
Two cylinders (Fig. 85) and their covers,
Two steam-chests,
Two slide-valves,
Two cross-heads,
Six axle-blocks (Fig. 87),
Two parallel bar supports (Fig. 88),
Four eccentrics,
Four eccentric rods and bands,

Slide-Valve Locomotive Engine.

One reversing quadrant (Figs. 89 and 90),
One starting lever,
One starting quadrant,
One steam-tap casting (Fig. 92),
Two name-plates,
Man-hole and plug,
Steam-tap, stuffing-box, and gland,
Four buffers (Fig. 93),
Lower part of chimney (Fig. 94),
Top of chimney,
Chimney-tube,
Steam-dome,
Pair of leading-wheels,
Two pair of driving-wheels,
T-joint for steam-pipe.

THE BOILER.—This is made with six $\frac{5}{8}$-inch tubes. The two ends should be clamped together and drilled for the tubes, two below and four immediately above them, all six being kept as low down in the boiler as possible. The tubes, which will be rather longer than the boiler, must be placed in position in the front and back of the latter, and their ends spread well out and hammered over, after which a little solder is to be run round the end of each tube to close the joint.

The front end of the boiler will consist of a flanged circular plate, but at the foot-plate end the lower part of the boiler is cut away for a length of

$2\frac{3}{4}$ inches to form the fire-box, and the back plate must be of a shape to fit this opening. A plate 4 inches wide and 7 inches long will allow of a sufficiently wide flange; it must be bent twice at right angles in opposite directions to fit the boiler, and the top and bottom must then be rounded and the piece flanged all round.

The ends with the tubes *in situ* are now to be placed in the body of the boiler, and it will be best to rivet in the fire-box end even if the front end is only soldered. After the rivetting is finished, solder should be run into the joints. Further details respecting flanging and rivetting will be found in Chap. X.

The dome should be turned up, and the valve fitted to the top of it. [As to the fitting of this latter, see Chap. XI.] It may then be soldered on to the boiler, but before this is done, the steam-tap (Fig. 92) must be put in. Directions for making this part and the rest of the boiler-fittings will also be found in Chap. XI. The inlet pipe from the dome should be supported by a strip of metal fastened across the dome-hole in the boiler.

Any firm supplying castings will also supply the boiler already finished with steam-tap and other fittings, and I certainly recommend the purchase of it in this complete form to all but the most experienced amateurs.

THE SMOKE- AND FIRE-BOXES.—The ends of the smoke- and fire-boxes are flanged, and the sides of both parts are made of sheet brass rivetted or brazed inside the flanges of the ends. One end of the smoke-box, as also one end of the fire-box, has a hole in it the size of the boiler, and the two ends of this latter are supported in these holes; the other end of the smoke-box has a rather smaller hole to which the smoke-box door is fitted, while the other end of the fire-box has a small hole to be closed by the furnace door (Fig. 95). The lower part of the fire-box is left open for the lamp.

THE FURNACE DOOR.—The half-hinge seen at the lower part of this piece in Fig. 95 must be cut off, and after it has been drilled to take the hinge-pin and three rivets, it is to be rivetted on to the back of the fire-box. The door itself is to be polished up, and the hinge drilled and attached to the half-hinge on the fire-box by means of a pin run through both parts. The door may be finished off with a small knob or handle attached to a latch.

THE SMOKE-BOX DOOR is formed from a piece of tin or thin brass, which is either spun in the lathe or hammered into shape. [Directions for spinning will be found in the chapter on Boiler-Making.] The form of the handle seen in Figs. 96 and 98 will be familiar to all who have noticed the front of a real locomotive. On the inside of the door this handle works the latch; a piece of iron about half an inch wide is fastened across the front of the smoke-box, and the latch catches into a slot cut in the iron.

When finished, the parts just described should measure:—

The smoke-box 1¾ in. long × 4 in. diameter.
The fire-box 3 ,, ,, × 3¾ ,, wide and 5 in. deep.
The boiler 8¾ ,, ,, × 3½ ,, diameter.

THE FUNNEL.—The base (Fig. 94) and the top of the funnel should be brazed to the funnel tube, and the funnel is then to be rivetted, bolted, or brazed to the top of the smoke-box, in which of course a hole must be made rather smaller than the diameter of the lower part of the funnel.

The fire- and smoke-boxes are to be bolted to the bed-plate later on by means of small angle-plates.

We may now place the boiler and the smoke- and fire-boxes on one side, and proceed to the mechanical portion of the engine. In the working drawings, two of which (Figs. 96 and 98) are on a scale of 3 inches to the foot, and one (Fig. 97) on a scale of 4 inches to the foot, the same letters are applied to the same parts in each drawing. Fig. 98 shows the engine in elevation, Fig. 96 in section, and Fig. 97 shows the working parts as seen from underneath: A is the fire-box; B, the boiler; C C, the tubes, of which there are six; D, the smoke-box; E, the smoke-box door; F, the steam-tap; and G, the handle to the same.

The form of the bed-plate will be seen in Figs. 96,

97, and 98 ; the front or leading pair of wheels run under the bed-plate, but for the larger wheels there are four slots in the plate, as seen in Fig. 97.

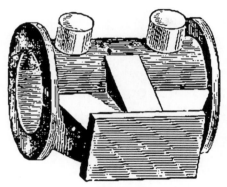

Fig. 85.

THE CYLINDER (Fig. 85) has two lugs cast upon it which are to be drilled and tapped for screws. These screws fasten the cylinder firmly to the bed-plate, passing through the latter from above. The cylinder

Fig. 86.

cover (Fig. 86) will be found somewhat different in construction from those which have been previously described, a piece being cast on it which serves to support the guide-bars. While the cover is in the lathe, a hole for the piston-rod must be bored, and where this passes through the oblong casting on the top of the cover, it must be enlarged and threaded to form the stuffing-box.

The castings for the slide-valve and chest, and for the glands of the stuffing-boxes, are similar in all respects to those for the engines already described.

The two steam-chests are joined by the T-piece included in the castings; the steam-pipe from the boiler is joined to the long arm of the T, and the two short arms are bored and screwed on the inside, as are also the steam-chests; then for each side a short piece of pipe is to have a screw cut upon it throughout its length, and to have two small nuts fitted upon it. This pipe must first be screwed into the arm of the T-joint for half an inch, then the other end is to be screwed into the steam-chest, but not so far as to withdraw it entirely from the T-joint or jamb the slide-valve. The two nuts are then screwed up against the T-piece and the steam-chest respectively, with a little packing of hemp between.

Fig. 87.

Fig. 87 shows the form of the casting for the crosshead, and also for the axle-blocks, and Fig. 88 shows the support for the back ends of the guide-bars; these latter may be screwed into the top of the cylinder-cover, and the support drilled for the other end; but a better method is to file up the guide-bars from square pieces of iron or steel 3 inches in length, as seen in the engine represented on the cover of this

FIG. 88.

book. Each bar must of course be perfectly straight and flat upon its inner or guiding surface, but the opposite surface—that is, the top of the upper bar and the bottom of the lower bar—may taper towards each end, the bar being $\frac{3}{16}$ of an inch thick in the middle, and $\frac{1}{8}$ of an inch thick at either end, the width being $\frac{3}{16}$ of an inch throughout. If the bars are made in this way, they must be drilled at each end and screwed to the ends of the projections on the cylinder-cover and the guide-bar support; and the length of these projections, coming between the guide-bars and thus determining their distance apart, must of course be carefully matched, care being also taken that the middle point in their length coincides with the centre of the piston-rod. The guide-bar support is shown *in situ* at H (Fig. 98).

THE CROSS-HEAD is to be fitted to the guides and piston-rod, and the top drilled to take the pin by which the connecting-rod is attached to it.

THE CONNECTING-ROD is a forging and is forked. At its top end it works upon a pin screwed into the driving-wheel; and outside the connecting-rod, the coupling-rod works upon the same pin, both rods being kept in place by the head of the pin. The connecting-rod measures $3\frac{1}{2}$ inches in length from centre to centre.

THE COUPLING-BAR is made of steel, and should be $\frac{1}{4}$-inch wide by $\frac{1}{8}$-inch thick, with a head $\frac{3}{8}$-inch long at each end, measuring $\frac{3}{8} \times \frac{1}{4}$-inch; that is to say, a piece of flat steel $\frac{3}{8} \times \frac{1}{4} \times 5\frac{1}{2}$ is to be reduced $\frac{1}{16}$ of an inch upon each surface, except the $\frac{3}{8}$-inch at each end. The coupling-bar may be bent as shown in Fig. 97, but it is preferable to keep it straight, and either place a distance piece over the pin of the trailing wheel, or make the pin with a shoulder against which the connecting-rod will work.

The pin by which the cross-head and connecting-rod are connected should be $\frac{3}{16}$ of an inch in diameter, with a $\frac{1}{4}$-inch head at one end, and a $\frac{1}{8}$-inch screw at the other. If a corresponding thread is cut on one side of the fork of the connecting-rod, a nut will not be necessary.

The driving-pin should be of the same diameter, but the two pins will of course not be of the same length.

THE AXLE-BLOCKS.—These must be filed up to fit the spaces provided for them in the bed-plate; the lower edge of this latter must be drilled, and a $\frac{3}{32}$-inch pin inserted in the centre of each axle-block space; a corresponding hole is to be drilled in the top of each axle-block, and when the engine is completed a spiral spring must be slipped over each of these pins, and these bearing upon the top of the axle-blocks will take the weight of the engine.

A piece of metal is screwed on across each of the openings for the axle-blocks, to prevent these latter from falling out when the engine is lifted up. The blocks must be drilled for the wheel-axles, which latter should be cut from $\frac{3}{8}$-inch steel bar.

At 1 (Fig. 96) is seen the guide to the slide-valve rod; this is a small casting screwed on to the bed-plate as shown.

The wheels will of course be turned up and keyed on as usual, great care being taken that the two wheels of each pair are of exactly the same diameter.

There are two eccentrics to each valve, each eccentric strap and rod being cast in one piece. The eccentrics are to be centred and turned up and bored, and the straps fitted to them as already described in former chapters; but at this point we come to an important difference, for the engines hitherto described have been constructed for forward motion only, whereas the engine now under consideration has reversing gear, which not only enables the driver to run the engine the reverse way, but also enables him to give the slide-valve any desired shortness of stroke, thus admitting steam to the cylinder for only part instead of for nearly the whole of the stroke. This is technically called *working expansively;* sometimes another valve, called an "expansion valve," is added to large engines, in addition to the slide-valve, for the purpose of cutting off the steam at an earlier period in the stroke of the piston than could conveniently be done with the slide-valve alone. However, to return to construction.

REVERSING GEAR.—Instead of being attached direct to the slide-valve rod, the eccentric-rods are attached to either end of a link. The links, which are cut out from flat pieces of iron, must be marked out with a pair of compasses, taking the distance from centre of shaft to the end of eccentric-rod as the radius for the curve of the slot. A number of holes are then drilled where the slot is to be, and the links are roughly cut out, leaving a lug at top and bottom for the attachment of the eccentric-rods, and another at the extreme lower end for the attachment of the lifting rod, as seen at K (Fig. 96). The links should now be fastened together, and the slots filed out, and both finished off before they are separated. Their length when finished should be $1\frac{3}{4}$ inch, their width in the centre $\frac{3}{8}$ of an inch; the slot should be $1\frac{1}{4}$ inch long, by $\frac{1}{8}$-inch wide, and the distance between the centres of attachment of the eccentric-rods should be $\frac{7}{8}$ of an inch. The valve-rod head is really a short fork which clasps the outer side of the link, and is bolted to a piece of brass which slides in the slot of the latter.

The rocking bar, L, will present no difficulty. It may be made to work between the points of two screws, screwed through the bed-plate on either side, or in small brass blocks screwed to the inner side of the bed-plate. On the rocking bar are three levers, to two of which the lifting rods are attached; and to the centre one is jointed the reversing-rod, a flat bar, $\frac{3}{16}$-inch wide, $\frac{3}{32}$-inch thick, seen at M. In Fig. 96 the rocking bar itself is at the end of the lever nearest to the letter L. The levers are fixed at this point; the other joints shown are hinge-joints.

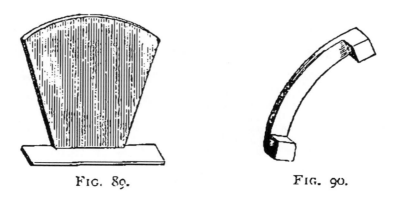

FIG. 89.   FIG. 90.

Figs. 89 and 90 show the quadrant in which the reversing lever works; the lever itself being shown at Fig. 91. As will be seen, it consists of the main lever, $\frac{1}{4}$-inch wide by $\frac{3}{16}$-inch thick, and a catch-bar lifted by a small handle at the top. The handle is bent as shown, and is attached to the catch-bar by a short link, the bar working in two bands which are rivetted on the reversing lever. The catch-bar

FIG. 91.

is kept down in the niches in the quadrant by the spring at the top. Another quadrant with longer legs is supplied for the starting lever; this should be engraved with the words "on" and "off," either by means of a graver, if the worker have sufficient skill in the use of this tool, or etched in with acid.

The starting lever is also supplied as a casting, but a small iron or brass handle must be turned for it. The starting lever is shown at G (Fig. 98); the quadrant has been omitted from the drawing.

There are two name-plates which may be soldered on to the boiler, see Fig. 98.

Fig. 93 shows one of the buffers; the rounded portion is to be cut off, and the pillar then left at the back of the square piece should be threaded, so that

FIG. 92.

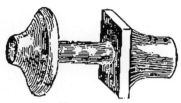

FIG. 93.

the buffer when finished can be screwed into the buffer beam. A hole $\frac{3}{16}$ of an inch in diameter must be bored right through the portion of the buffer last mentioned; that is to say, from the front through the

FIG. 94.

screw at the back. This hole is to be enlarged to $\frac{3}{8}$ of an inch for the first half-inch of its depth. The front portion of the buffer must be turned and made to slide one quarter of an inch into the $\frac{3}{8}$-inch hole in the back; into the back of the front part of the buffer is screwed a piece of $\frac{3}{16}$-inch steel $1\frac{1}{2}$ inch long; round this a spiral spring $\frac{1}{2}$-inch long is put immediately behind the buffer, and the two parts being put together, a small nut is screwed on to the end of the piece of $\frac{3}{16}$-inch steel projecting at the back of the buffer box. This part is now finished.

The "cow-catchers" are made from pieces of flat bar iron, and are bolted to the front of the bed-plate, a recess being cut for them in the buffer beam which

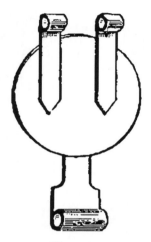

FIG. 95.

is screwed on over them. They will measure $\frac{1}{8}$-inch thick and $\frac{3}{8}$-inch wide at the top, tapering to $\frac{1}{8}$-inch at the lower end.

The driving-wheels are provided with covers or boxes; the trailing-wheels have no boxes, being covered by the tanks at the side of the foot-plates. These tanks, together with the side-plates, are cut out of stout tin-plate; and where there is so much heat, rivets are better than solder for joining the edges of the pieces from which the tanks are to be formed, to the side-plates. A bright steel hand-rail supported by small brass pillars screwed into the boiler will add to the finished appearance of the engine.

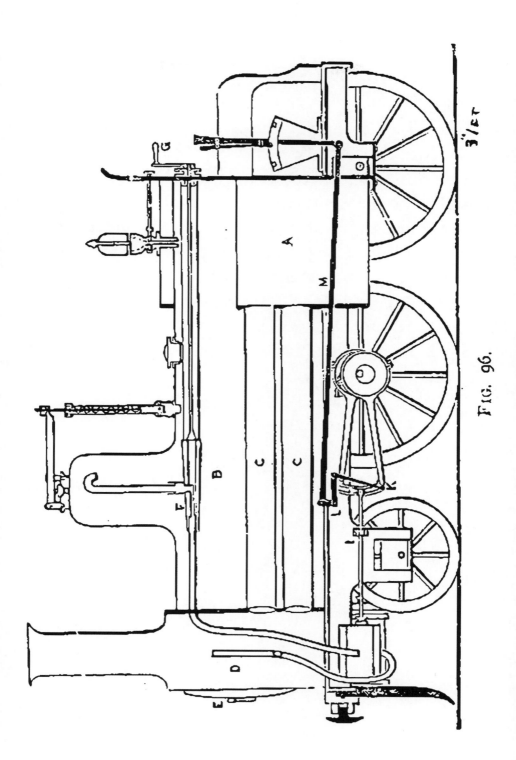

Fig. 96.

The spring seen over the leading wheels is only a sham one, and may either be made from separate pieces of spring steel bound together by a brass strap, or in one solid piece. In putting the engine together, the exhaust-pipes from the cylinders must be brought up through the bottom of the smoke-box, and cut off just below the funnel.

The lamp is in the form of an oblong box, the shape of the foot-plate of the engine, under which it hangs; six curved brass wick tubes project from it into the fire-box.

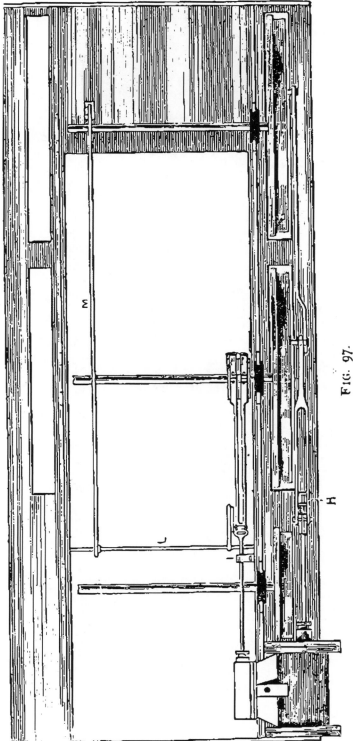

Fig. 97.

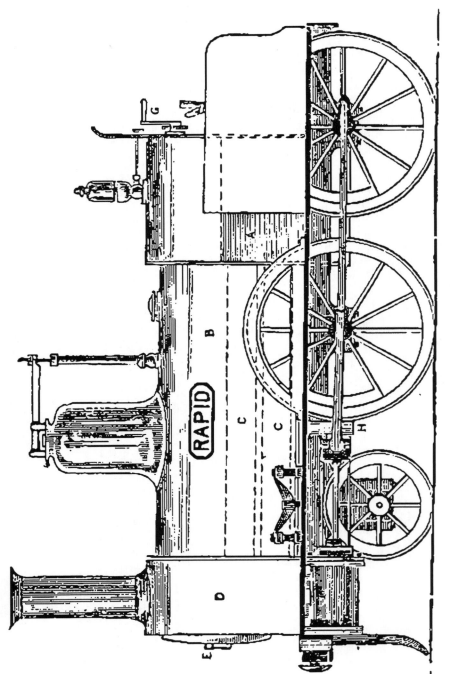

Fig. 98.

CHAPTER X.

BOILER-MAKING.

ALTHOUGH I do not altogether recommend amateurs to make their own boilers, there is no reason why they should not do so, provided they get them properly tested by a trustworthy firm. Boilers are tested by hydraulic pressure, and a pressure-gauge and force-pump are the necessary apparatus. I need hardly say that no boiler, whether made at home or bought, should be used without a safety-valve duly proportioned to the size of the boiler and its proper working pressure.

There are three ways of making a model boiler: (i.) soft soldering; (ii.) brazing, or, as it is sometimes called, hard soldering; and (iii.) rivetting.

The first method will answer well enough in cases where the pressure required is very slight, as, for instance, in engines with single-action oscillating cylinders.

Brazing very frequently proves a stumbling-block in the path of the amateur. The solder, usually called spelter, is mixed with borax and water, and spread along the inside of the joint. A forge is required, and with a clear fire and just the right heat,—above all, with plenty of practice,—a good joint will be the result. I cannot, however, recommend the process to any but a practised hand.

Rivetting is far easier than would appear at first sight, and this is the process I would recommend to those who *must* try their hand at boiler-making. Let us, therefore, assume this to be the method adopted by the reader.

The tools required are not many: a pair of metal shears for cutting thin sheets of metal, or a cold chisel for thicker sheets; a light hammer and a rivet-heading punch. This latter tool is for forming the heads of the rivets; it is made from a piece of steel about $\frac{1}{4}$ of an inch in diameter and 3 inches in length, which is centre-punched and slightly drilled with a, say, $\frac{1}{8}$-inch drill. The small conical hole thus made must be rounded either by finishing with a drill of the same section as the rivet heads are required to be, or by making the steel red-hot and hammering into the hole a piece of cold steel similarly rounded. The hole having thus been formed of the right shape, this end of the steel heading-punch is to be hardened in the usual way.

The most easily constructed boilers are those shown in Figs. 99 and 100. In the vertical boiler, shown in Fig. 99, the two ends are flanged either by hammering or by spinning in the lathe. The latter process will produce neater ends, and to effect it

Boiler-Making: 143

the disc of copper should be cut to the right size and placed between two discs of wood of the inside diameter of the flange. The whole is now to be put into the lathe, which is to be driven at a quick speed, while the edge of the copper is turned down with a burnisher. Should the metal appear to get thinner

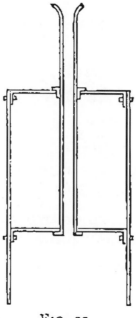

Fig. 99.

in any part, the burnisher must be worked towards that part to bring the metal back. The tool should be provided with a long handle, and the rest should have a series of holes along the top of it, in one of which is placed a pin to act as a fulcrum to the tool; and a little grease or soap and water should be used as a lubricant between the work and the tool.

By using a pattern of the shape required, the ends

may be spun up to a rounded form ; in this case the piece of wood between the back centre and the work must be of small diameter to allow the tool to be worked over the surface of the disc of metal. Should the metal slip, a little powdered resin may be put on the pattern. The metal will require to be annealed once or twice in the course of spinning it up. Only thin metal for small boilers can be spun up in this manner.

The body of the boiler may be formed from sheet copper or from a piece of copper tubing. In case the former is chosen, the sheet must be bent round ; its two edges, overlapping $\frac{3}{8}$ to $\frac{1}{2}$ an inch, should be drilled at each end for a rivet which may be put in and finished off at one end, but at the other end it should have only a tap or two with the hammer, as it is only meant to keep the two edges properly overlapping while the rest of the rivets are put in, and should therefore be finished off last of all. The rivets suitable for the smaller boilers are $\frac{1}{4}$ of an inch long, and rather over $\frac{1}{16}$ of an inch in diameter, with a flat head $\frac{3}{16}$ of an inch in diameter. The price of a pound of these rivets, containing several hundreds, is 2s. 6d. The larger-sized boilers will require rather larger rivets, which will cost somewhat less per pound.

For rivetting the body of the boiler, a stout bar of iron, rather longer than the boiler and supported at its two ends, will be necessary. It will be best to drill only two or three holes at a time, rivet them, and then drill two or three more, and so on. The rivets are put through from inside the boiler, which is then placed so that the flat heads of the rivets rest upon the iron bar; now give each rivet a tap or two with the hammer, and then, placing the heading-punch upon the rivet, give the punch a few blows, and a neatly rounded rivet head will be the result. The rivets should be placed from $\frac{3}{8}$ to $\frac{1}{2}$ an inch apart.

The top of the boiler must be drilled or punched for the chimney ring before it is rivetted to the body. For the rivetting of this part, the bar of iron must be held by one end in a vice.

Before the bottom is put in, the chimney may be soldered into it; as will be seen in Fig. 99, the lower end of the chimney is spread out or flanged, and is soldered to the bottom of the boiler with ordinary plumbers' solder. The chimney should be of copper or brass tube, $\frac{1}{2}$ an inch in diameter and upwards, according to the size of the boiler. The lower end of the boiler is rivetted with the flange downwards, as shown in the figure.

The chimney ring is turned up from a casting and soldered to the top of the boiler and to the chimney, and this latter is finished off by soldering or rivetting on to the top of it an ornamental cap, either turned up from a casting or spun in the lathe.

Last of all, soft solder is to be sweated into all the

seams, and to do this a little chloride of zinc should be run along the seam, and a little solder dropped on; a blow-pipe flame is then applied to the seam, and the solder will run into it as the flame is moved along. Some care should be taken not to use too much solder, as any surplus on the outside of the boiler will have to be removed, if a neat finish is desired.

The lower part, or fire-box, may be cut away on two sides to allow the air free access to the lamp-flame, or the same object may be accomplished by rivetting three short iron legs to the fire-box.

The boiler shown in Fig. 100 is more easily

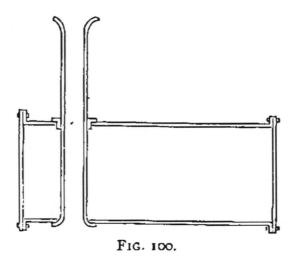

FIG. 100.

rivetted; but the body of the boiler must be flanged either by spinning or hammering, and it is not so easy to flange the body as it is to flange the two ends. Boiler tube already flanged may, however, be had from any copper worker's, and should cost about

1*s.* per lb. for the tube, and 2*s.* and upwards, according to size, for the two flanges. If copper tube be used for the body, the rivetting in this boiler will be all on the outside, so that perhaps this is the most simple boiler an amateur can take in hand. The chimney will, of course, be put in before the ends are attached.

Boilers of over 8 or 9 inches in length should always be stayed, especially if they are intended to carry more than a few pounds pressure. The stays may be made of $\frac{1}{8}$-inch wire reduced a little at each end for $\frac{1}{8}$ of an inch or so to form a shoulder. They must be placed across the boiler and rivetted over before the ends are rivetted on.

Either of these methods of construction may be used for either a horizontal or a vertical boiler. In the case of a horizontal boiler, the furnace must be separately made from thin sheet iron, or stout tin plate may be used.

Fig. 101 shows in section a simple form of marine boiler; as will be seen, the fire-box is made within the body of the boiler, a flanged plate being rivetted in the latter to form the bottom of the water space. The funnel end of the boiler and furnace may be completely closed, but at the other end the fire-box should be left open to admit the air to the lamp-flame.

In order to economize room by making the boiler more the shape of the boat, when it is to be put entirely under deck, marine boilers are often made with flat tops instead of round. In this case they are best made from sheet copper, and the bottom of the boiler should be rivetted in before the seam (which

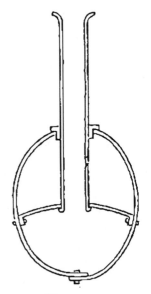

FIG. 101.

will come at the bottom of the fire-box) is closed up and rivetted. In boilers with flat tops it is better to put a few stays; these must also be rivetted over and soldered before the bottom seam is closed.

For large-sized models of marine and locomotive engines, tubular boilers are preferable. The tubes are usually brazed in; and so few amateurs can take this work in hand, that it is unnecessary to occupy space with further details.

The copper used for model boilers varies from No. 15, Birmingham wire gauge, for the larger sizes, down to No. 20 or 22 for the smallest. An easy rule is to divide the diameter of the boiler in inches by 100; this will give in decimals the thickness of the metal suitable for a pressure of 15 or 16 lbs. to the square inch, which is more than sufficient for any model.

The following are the minimum sizes of boilers suitable for the engines described in previous chapters:—

CYLINDER.	BOILER.

Horizontal or Vertical Engine:—

1 in. stroke,	$\frac{1}{2}$ in. bore	. . .	$2\frac{1}{2} \times 5\frac{1}{2}$ in.
$1\frac{1}{2}$,, ,,	$\frac{3}{4}$,, ,,	. . .	3×7 ,,
2 ,, ,,	1 ,, ,,	. . .	$3\frac{1}{2} \times 10$,,
$2\frac{1}{2}$,, ,,	$1\frac{1}{4}$,, ,,	. . .	5×13 ,,
3 ,, ,,	$1\frac{1}{2}$,, ,,	. . .	$6\frac{1}{2} \times 15$,,

Single Cylinder Slide-Valve Launch, or Double Oscillating Cylinder Marine Engine:—

$\frac{3}{4}$ in. stroke,	$\frac{1}{2}$ in. bore	. .	$6 \times 3\frac{1}{2} \times 4$ in.
1 ,, ,,	$\frac{3}{4}$,, ,,	. .	$7 \times 4 \times 4\frac{1}{2}$,,
$1\frac{1}{2}$,, ,,	1 ,, ,,	. .	$9 \times 5 \times 5\frac{1}{2}$,,

Double Cylinder Slide-Valve Launch Engine:—

$\frac{3}{4}$ in. stroke,	$\frac{1}{2}$ in. bore	. .	$9 \times 5 \times 5\frac{1}{2}$ in.
1 ,, ,,	$\frac{3}{4}$,, ,,	. .	$11 \times 6 \times 7$,,
$1\frac{1}{2}$,, ,,	1 ,, ,,	. .	$13 \times 7 \times 9$,,

It is not, however, advisable to use the minimum size as supplied at the shops; but for any given size of cylinder, the boiler given above for the size larger should be used, while if a horizontal or vertical engine of $2\frac{1}{2}$ or 3-inch stroke is required to do any real work—such as driving a sewing-machine, for instance—the boiler should be about 9 × 20 inches or larger, instead of $6\frac{1}{2}$ × 15. The minimum diameter for tubes in model engine boilers is half an inch.

LAMPS.—A lamp suitable for most model boilers is shown in Fig. 25, but for marine engines it is advisable to have the reservoir of spirit away from the heat of the boiler and furnace; the reservoir in this case may be above the level of the furnace. The tube leading to the furnace enters the reservoir close to the top, or it may be carried through the bottom of the reservoir nearly to the top. A wick from this tube hangs down into the spirit, and acts as a siphon in keeping the lamp supplied. The number of flames required will of course depend upon the size of the boiler and the amount of work expected from it.

CHAPTER XI.

BOILER AND OTHER FITTINGS.

THIS chapter will be devoted to a description of taps, valves, and similar odds and ends, which, although required in fitting up an engine, have found no place in the foregoing pages.

STEAM-TAPS.—Fig. 20, page 31, shows the casting for a steam-tap, although it may easily be turned out of any round piece of brass that is thick enough for the purpose. The piece should be first bored through from end to end; the drill used must depend upon the size of the tap. The piece may now be turned up between the lathe-centres and bored for the plug; after boring, the plug-hole should be made slightly conical with a rimer. The plug may be made in two ways; either it may be turned up as

FIG. 102.

shown in Fig. 102, a hole being bored through the top, in which the handle is afterwards fitted; or it may be formed from a piece of brass wire, which, after the plug has been ground in, is bent over and filed up square to form the handle, as shown in Fig. 103. In

Fig. 103.

either case, the plug having been turned to fit the hole in the tap, it is put into the lathe and ground into place with a little fine grindstone or coke dust, or pumice-stone powder and water, finishing off with a little putty powder.

When the plug has been fairly ground into place, the steam-way must be bored through it, and in small taps the plug may then be rivetted over at the lower end, or the plug hole in the tap may be slightly enlarged at the lower end, and the plug having been put in place may be filed off flush and then spread a little by punching in the centre with a centre-punch; but in larger taps it will be found best to drill and tap the lower end of the plug and put a screw into it, while in the largest size the plug may project below the tap and be fitted with a nut.

If a curved tap should be required, it is most easily made by inserting a piece of bent tube into a straight tap, reducing the latter at the junction so that the join may not be perceptible.

The steam-tap shown in the drawing of the locomotive engine (Fig. 96) is made in rather a different manner. The casting for the tap (Fig. 92) is bored with a conical hole, and a straight hole is drilled through the projection on one side; into this latter is soldered or screwed the pipe going up into the dome. To the small end of the conical casting a pipe leading into the smoke-box is attached, and this pipe is joined by means of a forked junction or T-piece to the two pipes leading to the steam-chests. To the other end of the cone another tube is attached; and fastened to the foot-plate end of this is a nut screwed on the inside, and fixed into the back-plate of the boiler, as seen in Fig. 96.

The plug for this tap is to be bored half-way up through its length and half-way through from one side, as seen in the illustration. Attached to the larger end of the plug is a long bar carrying at its other end the starting handle. The bar works through a nut, which screws into the one attached to the back plate of the boiler, and also screws up against a small collar which must be shrunk on to the bar.

It will now be seen that by screwing up the inner nut which works against the collar on the steam-tap rod, the plug of this latter is forced home and kept steam-tight, and also by unscrewing the nut the plug may be pulled out, should it at any time require attention.

The steam-tap is supported by a stay across the dome, not shown in the engraving.

STEAM-WHISTLES.—The ordinary steam-whistle, shown in Fig. 104, is simply a tap upon one end of

FIG. 104.

which a piece of closed brass tubing of suitable size has been soldered ; before the tube is soldered on, however, it must be partially blocked up at the open end by soldering in a short piece of round brass fitting the tube but having a flat filed upon one side. The position of this piece of brass is shown in the drawing, and the mouth of the whistle is made with a file just above the block and upon that side of the tube to which the flat part of the block is turned.

In the drawing the tap is shown turned off, and the handle is thus hidden behind the whistle. Unlike the handles of steam-taps, the handles of these steam-whistles should be at right angles to the whistle when turned on, otherwise the fingers may be brought into unpleasant proximity to the steam escaping from the mouth of the whistle when the latter is being turned on or off.

The bell whistle shown in the locomotive (Figs. 96 and 98) is given in section in an enlarged drawing in Fig. 105. As will be seen, it consists, besides the bell, of a tap with a cup at the top. Into this cup is screwed a piece of brass bearing a disc at the top, a hole is drilled a little way up this piece and another hole is drilled through it crossing the first one, thus forming a T-shaped steam-way with two outlets.

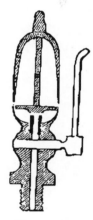

FIG. 105.

The disc almost fills the cup, leaving a very narrow opening all round; so that when the tap is turned on the steam fills the chamber below the disc, and escaping all round this latter, strikes against the edge of the bell, thus producing a shrill whistle.

UNIONS.—Fig. 106 shows a union. The nut and

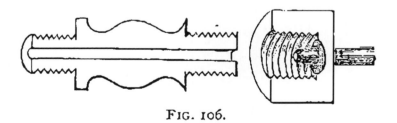

FIG. 106.

union-piece are cast in one; the former is to be cut off and bored out to within $\frac{3}{16}$ of an inch of one end, which is to be drilled to receive the pipe for which the junction is intended. The larger part of the union and the nut are to be screwed to fit one another, the union-piece being drilled, and, if desired, fitted as a tap. A collar must be brazed on about $\frac{1}{8}$ of an inch from the end of the tube with which the union is to be used.

SAFETY-VALVES.—Figs. 107 and 108 show the cast-

FIG. 107.

FIG. 108.

ings for a lever safety-valve. The main casting, shown in Fig. 107, must be turned up between the lathe-centres, and a hole drilled through it, say a $\frac{3}{16}$-inch hole for the size shown in the drawing. At the top end the edge of this hole must be taken off so as to leave a very short conical bearing-surface.

To make the valve, take a piece of brass about $\frac{1}{8}$ of an inch larger than the hole drilled in the casting; and turn $\frac{1}{4}$ to $\frac{3}{8}$ of an inch of it down to fit this latter exactly; then turn down the next $\frac{1}{8}$ of an inch to nearly fit the conical bearing in the casting, but it is to be noted that the surfaces actually in contact must be something under $\frac{1}{16}$ of an inch in width. The valve is now to be ground into its seat in the casting with a little grindstone dust and water, and three flats are then to be filed upon the spindle which has been left below the valve. These flats will form the steam-ways, while the three rounded edges left between them, fitting the central hole in the casting, will cause the valve to fall back truly into its place after it has been lifted. The top of the valve may now have a groove filed in it, as shown in Fig. 109 at A, which is an enlarged drawing of this piece.

Another piece of brass must be turned up and slotted and screwed into the rim of Fig. 107; and a piece of flat steel bar, drilled at the end, is to be pivoted in the slot of this last piece by a pin, rivet, or screw.

The weight (Fig. 108) must be turned up, the top end being held in a grip-chuck, and the other end supported by the back centre, or it may be turned up between the two centres. The top of the weight must have a slot cut through it, so that it may slide on the valve lever; two or three holes should first be drilled through, and the slot afterwards finished by filing. The valve piece must be screwed at the bottom, and it will then be finished and appear as shown in Fig. 109.

Several variations may be made in this valve; for instance, instead of making a guiding groove in the top of the valve piece, this latter may be filed off flat, and another piece of brass with a guiding slot in it

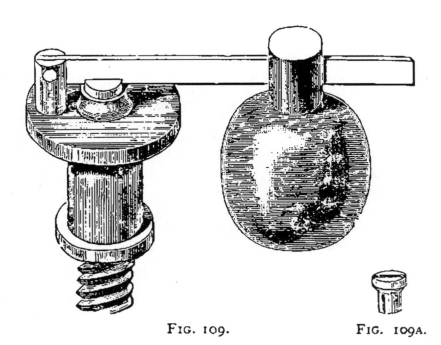

FIG. 109.   FIG. 109A.

may be screwed into the rim of the main casting in front of the valve; also the lever may be a bent one attached to a projection upon the rim of the main casting.

The pressure at which a lever safety-valve will let off steam is approximately found by multiplying the weight in pounds by the length of the lever from its point of support to the point at which the weight is suspended, and dividing by the area of the valve in inches multiplied by the distance between point of suspension of the lever and centre of the valve; the result giving the pressure in pounds per square inch. To get the exact pressure, an allowance must also be made for the weight of the lever and of the valve itself.

In full-sized engines the area of the safety-valve is made ·006 the area of the fire-grate, and the mitre is made not exceeding $\frac{1}{16}$ of an inch.

Spring-valves may be made either with or without a lever. If made with a lever, the valve itself is constructed in the same manner as the valve already described, excepting that the free end of the lever is forged out flat and pierced with a hole or slot. The spring acts upon a piston working in a tube, the upper end of the piston-rod being screwed and provided with a nut. If the end of the lever is not slotted, the spring-barrel must be pivoted at its lower end. The whole arrangement will be made quite clear by a reference to Figs. 96 and 98.

In the case of a spring-valve without a lever, the spring will act directly upon the valve, the escape of steam taking place through the tube containing the spring. Such a valve may also be made adjustable by placing a piston upon the top of the spring, a nut kept down by the cap of the spring-barrel being used to screw the piston-rod up and down. Fig. 110A shows the cap and nut in an arrangement of this sort; the lower part and centre of the nut are in one piece, but the top is filed up square, and the milled head by which it is worked is made with a square hole to fit the top of the nut.

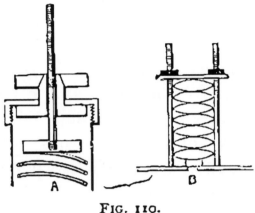

FIG. 110.

A variation of this last valve is shown at B in the same figure, where several pieces of watch-spring are substituted for the spiral spring.

By cutting a slot down the spring-barrel and attaching a pointer to the piston, either of these forms of valve may be made to show the pressure at which they are set.

Valves with weights are sometimes used for stationary engines, but some form of spring-valve is invariably used for locomotive and marine boilers, as the oscillations and vibrations to which such boilers are subjected would interfere with the action of the weight.

GOVERNORS.—Figs. 111 and 112 show two forms of governors; the first adapted to small models and the second suitable for small workshop engines. The castings for Fig. 112 cost from 2s. and upwards, according to size. Fig. 112 is half-size, and the position of the throttle-valve in relation to the governor is regulated by the small set-screw seen at A.

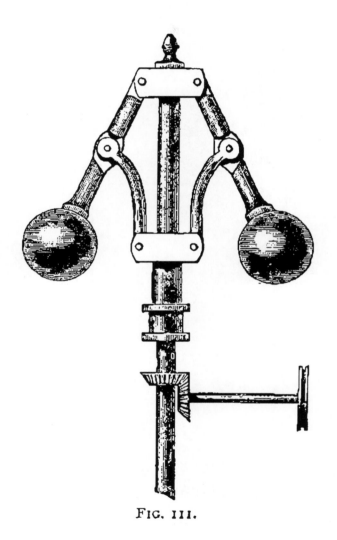

Fig. 111.

In Fig. 111 the forked lever, links, and throttle-valve have been omitted, being precisely similar to those shown in Fig. 112. The method of construction is made so clear in the drawings that no further description will be necessary.

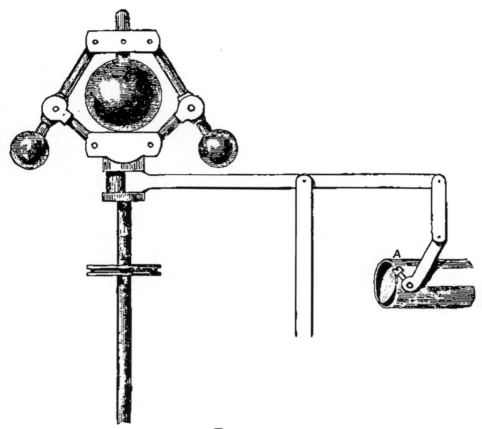

FIG. 112.

FORCE-PUMP.—Fig. 113 shows in section a small force-pump for supplying the boiler. It is worked by an eccentric on the main shaft of the engine. The barrel and valve-box of the pump are all in one casting.

The barrel is first bored out, then a hole is bored through the valve-box of the same size as the intake pipe. This hole must be enlarged for about two-thirds of the way down for the lower valve, and again enlarged for the first third of its length for the upper valve. The outlet hole is also to be bored; the valves should be made in the same manner as the

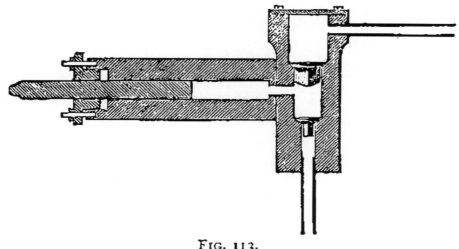

FIG. 113.

safety-valve already described, and must be ground into place while they form part of a long piece of metal, and afterwards cut off. It will be noticed that the lower part of the top valve must be the same size as the largest part of the lower valve. A plate is screwed on to the top of the valve-box, and the plunger barrel is fitted with a gland as shown in the drawing.

STEAM PRESSURE-GAUGES.—The ordinary Bourdon gauge is beyond the powers of an amateur; it consists of a flattened tube which is constructed by special machinery. It is curved, and firmly fixed at one end, the free end being connected by a series of levers with the index hand; the interior of the flat tube is connected with a siphon containing water, and this in turn is connected with the boiler. The tendency of the curved tube is to straighten itself in proportion to the pressure within it; the extent of this movement, and consequently the amount of the pressure, is shown by the index hand.

Another form of pressure-gauge, however, which lies well within the capabilities of an amateur, has been twice illustrated in the *English Mechanic*, and is shown in Fig. 114. It consists of two small

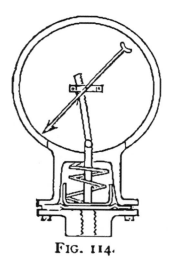

FIG. 114.

castings, the upper one bored out true, and the lower one having a flat surface and a screwed hole for attaching the siphon pipe. In the upper casting a tubular piston works against the pressure of a spiral spring; connected to the piston-rod by a hinge joint is another rod with a short rack at the top, which is kept by a spring against a pinion carrying an index hand. The upper casting also carries a ring into the back of which a dial is fixed, the front being closed with glass. Before the two castings are screwed together, a sheet of rubber is placed between them to prevent the water from the siphon tube being forced beyond the piston. The gauge is graduated by comparison with a standard instrument.

LOOSE ECCENTRIC.—One way of making an engine reversible has already been described in Chap. IX. An easier method, and one which may be applied to any engine, is by means of a loose eccentric. The eccentric itself is of the usual form, but it is not fixed upon the shaft, being kept in place on one side by a small collar on the shaft, while on the other side it works against a collar of the shape shown in Fig. 115. As will be apparent from this

FIG. 115.

drawing of the collar, half the thickness of rather more than half its surface (half the surface plus twice the angle of lead) is cut away, and the collar itself is fixed in its proper position on the shaft by means of a set-screw. Its abutments catch a short pin screwed into the side of the eccentric, and the collar thus carries the eccentric with it, and the engine will run whichever way it is started.

TAPPETS.—There are other ways of moving the slide-valve, as well as by means of an eccentric; for instance, an arm projecting from the piston-rod head and forked at its free end may be made to slide up and down the valve rod, but without touching it. Two small collars are then fixed to the slide-valve rod by means of set screws, so that the fork of the travelling arm coming into contact with one or other of these collars knocks the slide-valve up or down, as the case may be. The part of the slide-valve rod which carries these tappet collars must not however be connected directly with the valve, but a rocking-bar must be interposed, carrying a short lever on each side to which the two parts of the rod are attached; otherwise the valve will cover the upper port of the cylinder when the piston is at the top of its stroke, whereas the upper port is at that time required to be open, and *vice versâ*.

Another method of moving the slide-valve is by means of a cam on the shaft; in this case, the slide-valve rod has a slotted head, and at each end of this head and at right angles to its length is a plate of metal; the shaft passes through the slotted head and the cam acts first on one plate, then on the other, thus pushing the valve alternately up and down. The motion of a valve moved by a cam will vary from the sudden movement of the tappet action to the gradual motion of the eccentric, according to the shape of the cam, this latter being in fact an exaggerated form of the ordinary eccentric, the plates on the valve-rod taking the place of the eccentric strap.

In turning up some of the small pieces described in this chapter, it will be often found useful to have a brass or gun-metal chuck with a flat face to which the part requiring to be turned may be attached by means of a little solder; this method of chucking will also be convenient for cylinder-covers and other parts having flat surfaces by which they may be attached.

CHAPTER XII.

NOTES.

THE following notes may prove useful to some of my readers.

BURNISHING.—This process gives a highly finished appearance to all brass articles. Burnishers of various shapes may be purchased at the tool shops at from 6*d.* to 2*s.* 6*d.* each ;—half-round, flat, and oval curved are perhaps the most useful shapes for the amateur.

Previous to burnishing, the article should, if possible, be annealed by heating it and quenching it in water ; it must then be pickled in dilute nitric acid, washed in clean water, and dried in hot boxwood sawdust without the parts to be burnished being touched by the hands.

The burnisher must now be worked over all the parts that are to be burnished, using a little stale ale as a lubricant; some workers use soap and water instead of the ale, and others prefer ox-gall with cream of tartar and water. When the burnishing is finished, wash again and dry in boxwood dust, then lacquer.

LACQUERING.—The articles to be lacquered should,

if not burnished, be pickled in caustic soda or potash and water, to remove all grease, washed and dipped for a moment in nitric acid, washed again and dried in boxwood dust. Now make the pieces as hot as the hand can comfortably bear, and give each a thin coat of the lacquer with a brush.

Lacquer may be bought almost as cheaply as it can be made; but for the information of those who prefer to make their own, I here give a few recipes.

To one pint of alcohol or methylated spirit add one ounce of good pale shellac; this will form the body of the lacquer, and may be coloured to taste as follows:—

For a pale lacquer, add 3 parts of Cape aloes and 1 part turmeric.

For a golden lacquer, add 4 parts of dragon's blood and 1 part of turmeric.

For red lacquer, add 32 parts of annatto and 8 parts of dragon's blood.

Another method is to make saturated solutions of each of the above-named colours in spirit, and a small quantity of lacquer of any desired colour may then be easily made by adding more or less of any of the solutions to a little of the varnish.

PAINTING.—Enamel paints for painting the cylinders, etc., of engines may be bought at the model-shops, or may be made by adding copal varnish to ordinary paint.

SOLDERS.—A good soft solder may be made of

equal parts of lead and tin; or a finer solder may be made by doubling the quantity of tin.

Hard solder, or spelter for brazing, if required to stand a high temperature, should contain three parts of copper to one of zinc; and these proportions may be varied down to equal parts of each metal for ordinary work, while a still softer solder for brazing may be made by adding one part of tin to four of copper and three of zinc.

Silver solder, two parts silver to one of brass (ordinary pins), will be found useful for brazing small work.

FLUXES FOR SOLDERING. — For brass, copper, tinned iron, or zinc: chloride of zinc, made by dropping pieces of zinc into hydrochloric acid (spirits of salts) until there is no further effervescence. The fumes which arise during the process should not be inhaled. For tin or copper: rosin; and for lead: tallow or rosin may be used.

For brazing, the flux borax should be made into a paste with water and mixed with the spelter.

SETTING SLIDE-VALVES.—The eccentric-rod should first be adjusted to such a length that when the eccentric is at its farthest from the cylinder, the lower port is opened to exactly the same extent as the upper port is when the eccentric is at its nearest to the cylinder. After the length of the eccentric-rod has been thus adjusted, the eccentric must be fixed to the shaft in such a position that the upper or lower steam-ports will be just open when the piston is at the top or bottom of the cylinder.

The following explanation of one or two terms of frequent occurrence in descriptions of engines and their efficiency may also prove useful.

LAP.—The lap of the slide is the extent to which the valve overlaps the edges of the ports which it covers. If the face of the slide-valve exactly corresponded with the width of the steam-ports, steam would be admitted to the cylinder during the whole of each stroke, but by making the valve of such length that at the middle of its stroke it more than covers the ports, the steam is cut off before the stroke is finished, the rest of the work for that stroke being performed by the expansion of the steam already admitted to the cylinder.

LEAD.—This term is used to express the distance which the slide-valve stroke is in advance of the piston stroke. If the slide-valve were required to be in its middle position at the moment the piston is at the top or bottom of the cylinder, it is obvious that the eccentric would be set at right angles to the crank. It is found, however, that an engine works best when the ports begin to open for the next stroke just before the completion of the previous stroke; and in order to accomplish this, the eccentric is set at rather more than a right angle in front of the crank; and the difference between the angle at which the eccentric is set and that position which would bring the slide-valve to the middle of its stroke with the piston at the top or bottom of the cylinder, is called the *angle of lead*.

SUPERHEATED STEAM.—This is a term applied to steam which has been heated above the boiling point corresponding to its pressure. For instance, the normal temperature of steam under a pressure of 15 lbs. above the pressure of the atmosphere is 250° Fahrenheit. Now, if that steam be heated beyond that temperature while the pressure remains the same, or if it be maintained at that temperature while the pressure is reduced, as when it is expanding in the cylinder after the induction port is closed, it will be superheated.

UNIT OF HEAT.—This is the amount of heat necessary to raise one pound of water by one degree of Fahrenheit, and is represented mechanically by 772 foot-pounds.

One cubic inch of water will produce a cubic foot of steam.

HORSE-POWER.—In calculating the power of steam and other engines, 33,000 lbs. raised one foot in one minute is taken as equal to the power of one horse.

NOMINAL HORSE-POWER.—This is rather an indication of the size of an engine than of its power; in calculating the nominal horse-power of an engine the velocity of the piston is assumed to be 128 feet per minute multiplied by the cube root of the length of stroke in feet, and the steam pressure in the cylinder is assumed to be 7 lbs. per square inch, then the *nominal* horse-power equals $7 \times 128 \times \sqrt[3]{\text{stroke in feet}} \times \text{area of piston in square inches} \div 33,000$.

INDICATED OR ACTUAL HORSE-POWER.—This is calculated from the actual pressure in the cylinder and speed of the piston as follows:—Let A = area of piston in square inches, P = the average pressure of the steam in the cylinder in pounds per square inch, R = the number of revolutions per minute, and S = the length of stroke in feet, then the *actual* horse-power will be $\dfrac{2 A P R S}{33,000}$.

AVERAGE PRESSURE.— If an engine is working expansively, and the pressure in the boiler is 5 lbs. per square inch, and if the steam is cut off at half-stroke, the average pressure in the cylinder will be $4\frac{1}{3}$ lbs., and with a cut off at three-quarter stroke, $4\frac{4}{5}$ lbs. per square inch. The average pressure increases in the same ratio as the boiler pressure; thus, with 15 lbs. pressure in the boiler, the average pressure under the above circumstances will be about $12\frac{3}{5}$ and $14\frac{2}{5}$ lbs. per square inch.

ROUND STEEL for making drills, etc., is usually sold in one foot lengths. If ordered by the Birmingham wire gauge numbers, the name of the gauge must be specified, as it is also sold by what is known as the Lancashire gauge. This gauge runs from No. 1, which is about $\frac{1}{4}$ of an inch in diameter, to No. 60, which is about the size of a stout sewing needle. The price varies with the market; but supposing Nos. 1 to 12 to be 2s. 6d. per lb., then Nos. 13 to 18 would be 3s., and the price would increase about 6d. for every four to six numbers, going up to 7s. for Nos. 58 to 60.

SQUARE STEEL is made $\frac{1}{8}$ and $\frac{3}{16}$ inch square and $\frac{1}{8} \times \frac{1}{4}$ inch. With No. 1 round steel at 2s. 6d., this square hammered steel would cost 4s. per lb.

The following tables will be found useful for reference, but it is to be noted that authorities differ as to the exact weight, in some cases to the extent of 8 or 10 per cent., of wire and sheet metal:—

ROUND AND SQUARE WROUGHT-IRON WEIGHTS PER FOOT.

Side of Sq. or Diam.	Square Lbs.	Round Lbs.	Side of Sq. or Diam.	Square Lbs.	Round Lbs.
$\frac{1}{16}$	·013	·010	$1\frac{1}{16}$	3·763	2·955
$\frac{1}{8}$	·052	·041	$1\frac{1}{8}$	4·219	3·313
$\frac{3}{16}$	·117	·092	$1\frac{3}{16}$	4·701	3·692
$\frac{1}{4}$	·208	·164	$1\frac{1}{4}$	5·208	4·091
$\frac{5}{16}$	·326	·256	$1\frac{5}{16}$	5·742	4·510
$\frac{3}{8}$	·469	·368	$1\frac{3}{8}$	6·302	4·950
$\frac{7}{16}$	·638	·501	$1\frac{7}{16}$	6·888	5·190
$\frac{1}{2}$	·833	·654	$1\frac{1}{2}$	7·500	5·410
$\frac{9}{16}$	1·055	·828	$1\frac{9}{16}$	8·138	6·322
$\frac{5}{8}$	1·302	1·023	$1\frac{5}{8}$	8·802	6·913
$\frac{11}{16}$	1·576	1·237	$1\frac{11}{16}$	9·492	7·455
$\frac{3}{4}$	1·875	1·473	$1\frac{3}{4}$	10·21	8·018
$\frac{13}{16}$	2·201	1·728	$1\frac{13}{16}$	10·95	8·601
$\frac{7}{8}$	2·552	2·004	$1\frac{7}{8}$	11·72	9·204
$\frac{15}{16}$	2·930	2·301	$1\frac{15}{16}$	12·51	9·828
1	3·333	2·618	2	13·33	10·47

Table Showing the Weight of 100 Feet of Wire.

Descriptive Number.	Diameter in Parts of an Inch.	Diameter in Metric Millimetres.	Copper Wire in Lbs.	Brass Wire in Lbs.	Iron Wire in Lbs.
7/0	·500	12·700	76·576	72·006	65·45
6/0	·464	11·785	65·947	62·010	56·235
5/0	·432	10·793	57·104	53·752	48·858
4/0	·400	10·160	49·009	46·083	41·888
3/0	·370	9·449	42·388	39·858	36·229
2/0	·348	8·839	37·095	34·88	31·705
1/0	·324	8·229	32·155	30·235	27·482
1	·300	7·620	27·5445	25·922	23·562
2	·276	7·010	23·333	21·940	19·942
3	·252	6·401	19·451	18·290	16·598
4	·232	5·893	16·486	15·502	14·091
5	·212	5·385	13·768	12·948	11·766
6	·192	4·877	11·792	10·617	9·651
7	·176	4·470	9·4882	8·921	8·1095
8	·160	4·064	7·8414	7·373	6·702
9	·144	3·658	6·3516	5·972	5·4287
10	·128	3·251	5·0185	4·7189	4·2893
11	·116	2·946	4·1217	3·8756	3·5228
12	·104	2·642	3·313	3·1153	2·8316
13	·092	2·337	2·5926	2·4378	2·2158
14	·080	2·032	1·9603	1·8433	1·6755
15	·072	1·829	1·5879	1·4931	1·3572
16	·064	1·626	1·2546	1·1767	1·0723
17	·056	1·422	0·96058	0·9324	0·821
18	·048	1·219	·70573	·6636	·60319
19	·040	1·016	·49000	·46083	·41888
20	·036	0·914	·39698	·37328	·33929
21	·032	·813	·31366	·29493	·26808
22	·028	·711	·24014	·22529	·20525
23	·024	·610	·17643	·1659	·15079
24	·022	·559	·14826	·1394	·12671
25	·020	·508	·12252	·1152	·10472
26	·018	·457	·099243	·093318	·084824
27	·0164	·4166	·082384	·077466	·070414
28	·0148	·3759	·067093	·06308	·057345
29	·0136	·3454	·056654	·053273	·048422
30	·0124	·3150	·047097	·044286	·040255

Table Showing the Weight of One Square Foot of Rolled Metal.

Descriptive Number.	Thickness in Parts of an Inch.	Thickness in Metric Millimetres.	Copper in Lbs. (Sp. Gr. 8·986).	Brass in Lbs. (Sp. Gr. 8·448).	Iron in Lbs. (Sp. Gr. 7·680).
7/0	0·500	12·700	23·4	22·0	20·0
6/0	·464	11·785	21·715	20·416	18·56
5/0	·432	10·973	20·22	19·008	17·28
4/0	·400	10·160	18·72	17·6	16·0
3/0	·372	9·449	14·441	16·368	14·88
2/0	·348	8·839	16·285	15·312	13·92
1/0	·324	8·229	15·163	14·256	12·96
1	·300	7·620	14·04	13·2	12·0
2	·276	7·010	12·917	12·144	11·04
3	·252	6·401	11·794	11·088	10·08
4	·232	5·893	10·858	10·208	9·28
5	·212	5·385	9·9217	9·328	8·48
6	·192	4·877	8·9856	8·448	7·68
7	·176	4·470	8·2368	7·744	7·04
8	·160	4·064	7·488	7·04	6·4
9	·144	3·658	6·552	6·16	5·6
10	·128	3·251	5·9904	5·632	5·0965
11	·116	2·946	5·4288	5·104	4·68
12	·104	2·642	4·8672	4·576	4·16
13	·092	2·337	4·3056	4·048	3·68
14	·080	2·032	3·744	3·52	3·2
15	·072	1·829	3·3696	3·168	2·88
16	·064	1·626	2·9952	2·816	2·56
17	·056	1·422	2·6208	2·464	2·24
18	·048	1·219	2·2464	2·112	1·92
19	·040	1·016	1·872	1·76	1·6
20	·036	0·914	1·6149	1·584	1·44
21	·032	·813	1·4976	1·408	1·28
22	·028	·711	1·3104	1·232	1·12
23	·024	·610	1·1232	1·056	0·96
24	·022	·559	1·0296	0·968	·88
25	·020	·508	0·936	·88	·8
26	·018	·457	·8424	·792	·72
27	·0164	·4166	·76752	·7216	·656
28	·0148	·3759	·69264	·65118	·592
29	·0136	·3454	·63648	·5984	·544
30	·0124	·3150	·58032	·5456	·496

INDEX.

Axle blocks, 128, 131.

Bearings, 84, 95.
Bed-plate, 77, 92.
Boiler fittings, 151.
   „   Horizontal, 146.
   „   Locomotive, 33, 123.
   „   making, 141.
   „   Marine, 147.
   „   Size of, 149.
   „   Vertical, 143.
Boring bar, 54.
Brazing, 141.
Buffers, 93.
Burnishing, 168.

Castings, 8.
Centre punch, 19.
Connecting rod, 80.
Coupling bars, 130.
Cowcatchers, 136.
Crank, 28, 96.
   „   Double, 102, 116.
   „   pin, 78.
Crank-shaft, 77, 84, 98.
   „   „   Bent, 78.
Crosshead, 86, 96, 102, 128.

Cylinder, Boring, 52.
   „   Chucking, 46.
   „   covers, 16, 35, 62, 94, 127.
Cylinder, Locomotive, 127.
   „   Oscillating, 14, 34, 114.
Cylinder, Slide-valve, 46.

Eccentric, 77, 104.
   „   Adjusting, 88.
   „   Loose, 164.
   „   strap, 76.

Fire-box, 124.
Fluxes, 170.
Fly-wheel, 22, 88.
Force pump, 162.
Furnace door, 125.

Gauge, 163.
Glands, 68.
Governors, 160.
Guide bars, 83, 86, 96, 129.
   „   blocks, 86.

Horizontal engine, 42.

Index.

Horse power, 172.
   „   „   Actual, 173.
   „   „   Nominal, 172.

Lacquering, 169.
Lamp, 38, 150.
Lap, 171.
Launch engine, 99.
Lead, 171.
Links, 132.
Locomotive engine, 122.

Marine engine, 109.

Nuts, 17.

Oscillating cylinder engine, 12.

Paddle wheels, 118.
Painting, 169.
Piston, 16, 35, 70.
   „   Dimensions of, 73.
   „   rod head, 85.
Pressure, Average, 173.
   „   gauge, 163.
Pump, 162.

Quadrant, 133.

Reversing gear, 132.
   „   lever, 134.
Rivetting, 142, 145.
Rocking bar, 133.

Safety valves, 156.
Screws, 5, 18.
Screw-shaft, 104.

Slide valve, 73.
   „   „   rod, 86.
   „   „   Setting, 170.
Small power engines, 90.
Smoke box, 124.
   „   „   door, 125.
Solders, 169.
Spinning metal, 143.
Standards, 95.
Starting lever, 134.
Steam block, 17, 34, 111.
   „   chest, 67.
   „   ports, 58.
   „   taps, 151.
   „   ways, 24, 35, 111.
Steam whistles, 154.
Steel, round, 173.
   „   square, 174.
Stuffing-box (Cylinder), 66, 68.
   „   „   (Steam chest), 68.
   „   „   (Screw shaft), 104.
   „   „   glands, 68.
Super-heated steam, 172.

Tables, 174.
Tappets, 166.
Taps, Screw, 6.
   „   Steam, 151.
Tools, 2.

Unions, 154.
Unit of heat, 172.

Valves, 156.
Vertical engine, 91.

Whistles, 154.

MODEL STEAM ENGINES & BOILERS
OF ALL KINDS.

BRASS AND IRON CASTINGS, SETS OR PARTS.
(As supplied to the Author of this work.)

SEPARATE FINISHED PARTS.

ENGINE & BOILER FITTINGS
OF EVERY DESCRIPTION.

MESSRS. LUCAS & DAVIES, having had over 20 years practical experience in the manufacturing of the above, can confidently recommend their goods to Amateurs and others. The Patterns are of the newest designs (not copies of other makers) and carefully proportioned, so that the castings, which are of the best metal, make up easily, and with as little work as possible.

Mr. POCOCK, who has had castings from every firm in the trade, said, "Those supplied by us were superior to any others he had obtained, as to correctness of design, quality of metal, and ease of construction."

These Castings have been sent to all parts of the world, and have given universal satisfaction.

PRICE LIST, with Illustrations from photographs of the finished Engines, &c., post free, 4d.

LUCAS & DAVIES,
21, CHARLES STREET, HATTON GARDEN,
LONDON, E.C.

MODEL STEAM ENGINES,

CYLINDERS, PUMPS, STEAM & WATER GAUGES, SAFETY VALVES, ECCENTRICS, TAPS, ETC.,

Model Yachts and Steam Boats,

Blocks, Deadeyes, Skylight, Companions, Stanchions, Flags,
CANNON, ETC.

FITTINGS FOR MODEL SHIPS
(List Free).

MODEL SHIPS' FITTINGS MADE TO SCALE OR TRACING;
FIRST-CLASS WORKMANSHIP.

STEVENS' MODEL DOCKYARD,
22, ALDGATE, LONDON.

Send 3d. for Illustrated Catalogue.   100 Engravings.

ESTABLISHED 1843.

POSTANS & COMPY.,
Experimental Engineers
AND
SCIENTIFIC MODEL MAKERS,
75a QUEEN VICTORIA STREET, LONDON, E.C.
Factories: 25, Bread Street Hill, E.C.

Fourth Edition, numerous Cuts, Crown 8vo., Cloth.
2s. 6d.

THE DYNAMO: HOW MADE, And HOW USED.
By S. R. BOTTONE.

"Exceedingly plain, clear instructions for the manufacture of small dynamos."—*Journal of Science.*
"Gives minute instructions to amateurs, by following which they may be able to make, at trifling cost, a dynamo for themselves."—*Scotsman.*

SWAN SONNENSCHEIN & CO., PATERNOSTER SQUARE, E.C.

Printed in Great Britain
by Amazon.co.uk, Ltd.,
Marston Gate.